48

Advances in
Polymer Science

Fortschritte der Hochpolymeren-Forschung

W0230358

Light Scattering from Polymers

With Contributions by
W. Burchard and G. D. Patterson

With 74 Figures

Springer-Verlag
Berlin Heidelberg GmbH 1983

Editors

Prof. Hans-Joachim Cantow, Institut für Makromolekulare Chemie der Universität, Stefan-Meier-Str. 31, 7800 Freiburg i. Br., BRD

Prof. Gino Dall'Asta, SNIA VISCOSA – Centro Studi Chimico, Colleferro (Roma), Italia

Prof. Karel Dušek, Institute of Macromolecular Chemistry, Czechoslovak Academy of Sciences, 162 06 Prague 616, ČSSR

Prof. John D. Ferry, Department of Chemistry, The University of Wisconsin, Madison, Wisconsin 53706, U.S.A.

Prof. Hiroshi Fujita, Department of Macromolecular Science, Osaka University, Toyonaka, Osaka, Japan

Prof. Manfred Gordon, Department of Chemistry, University of Essex, Wivenhoe Park, Colchester C04 3 SQ, England

Prof. Joseph P. Kennedy, Institute of Polymer Science, The University of Akron, Akron, Ohio 44325, U.S.A.

Prof. Werner Kern, Institut für Organische Chemie der Universität, 6500 Mainz, BRD

Prof. Seizo Okamura, No. 24, Minami-Goshomachi, Okazaki, Sakyo-Ku, Kyoto 606, Japan

Prof. Charles G. Overberger, Department of Chemistry, The University of Michigan, Ann Arbor, Michigan 48 104, U.S.A.

Prof. Takeo Saegusa, Department of Synthetic Chemistry, Faculty of Engineering, Kyoto University, Kyoto, Japan

Prof. Günter Victor Schulz, Institut für Physikalische Chemie der Universität, 6500 Mainz, BRD

Dr. William P. Slichter, Chemical Physics Research Department, Bell Telephone Laboratories, Murray Hill, New Jersey 07 971, U.S.A.

Prof. John K. Stille, Department of Chemistry, Colorado State University, Fort Collins, Colorado 805 23, U.S.A.

ISBN 978-3-662-15740-4   ISBN 978-3-540-39484-6 (eBook)
DOI 10.1007/978-3-540-39484-6

Library of Congress Catalog Card Number 61-642

© Springer-Verlag Berlin Heidelberg 1983
Originally published by Springer-Verlag Berlin Heidelberg New York in 1983
Softcover reprint of the hardcover 1st edition 1983

2152/3140 – 5 4 3 2 1 0

Table of Contents

Static and Dynamic Light Scattering
from Branched Polymers and Biopolymers

Dedicated to Prof. Manfred Gordon on the Occasion of his 65th Birthday

Walther Burchard

Institute of Macromolecular Chemistry, Hermann-Staudinger-Haus, University of Freiburg, Federal Republic of Germany

The striking properties of synthetic polymers and biological macromolecules are largely determined by their shape and the internal mobility. Both quantities are closely related to the architecture of the molecules. This article deals with branched macromolecules in dilute solution, where the individual molecules are observed. The common technique for determining the shape of macromolecules is static light scattering. Information on the internal mobility and the translational motion of the mass centre can be obtained from the more recent technique of quasi-elastic or dynamic light scattering.

As a result of the mostly statistical mechanism of reaction, many different isomeric structures and a broad molecular weight distribution are obtained on polymerizing monomers with more than two functional groups. An interpretation of the quantities measured by the two light scattering techniques, i.e. the z-averages of the mean square radius of gyration $\langle S^2 \rangle_z$, of the particle scattering factor $P_z(q)$, of the translational diffusion coefficient D_z and of the reduced first cumulant Γ/q^2, as function of the weight average molecular weight M_w is not possible without a comparison with special well defined models.

Starting with simple regularly branched structures and ascending to the more involved randomly branched structures, the article presents various techniques for the calculation of the measurable quantities and concentrates on the polydispersity. The representation of the molecules by rooted trees is shown to be most adequate for an extension of the theory of regularly branched chains to randomly branched polymers where statistical means have to be applied. In the Flory-Stockmayer theory and the further developed cascade branching theory all average quantities which are measured by the two light scattering techniques, i.e. M_w, $\langle S^2 \rangle_z$, $P_z(q)$, D_z etc., are uniquely determined by the extents of reaction of the various functional groups which, statistically speaking, are link probabilities. The basis of the cascade branching theory and their rules for analytical calculations are displayed and elucidated with many examples.

The second part of the article compares theory and experiment. In general, good agreement is found with the behavior predicted by the cascade theory. An interpretation for the static and dynamic light scattering behavior is given, and new structure sensitive parameters are introduced by combination of the static with the dynamic light scattering data. In particular, the dimensionless parameter $\varrho = [\langle S^2 \rangle_z]^{1/2} \langle R_h^{-1} \rangle_z$ is shown to be a quantity of great relevance.

Finally, the applicability of the cascade theory to rather complicated systems with unequal functional groups, substitution effect, vulcanization of chains and long rang correlation as a result of directed chain reactions is shown. The limitation of the theory to essentially tree-like molecules and their unperturbed dimensions is outlined and the consequence of this error for the prediction of real systems is discussed.

Advances in Polymer Science 48
© Springer-Verlag Berlin Heidelberg 1983

A. Introduction

Polymer science is a fairly new field in the natural sciences. In the first 15 years it was certainly no more than a special subject of organic chemistry, but around 1940 this new topic started to spread its roots into all the traditional fields of science. The immense influence on biochemistry and biophysics is well known, and our present understanding of the molecular basis of biological processes is inconceivable without Staudinger's pioneering work[1, 2].

I) Equally eminent is the development of fundamental theoretical treatments in the decade 1940 to 1950 by W. Kuhn[3], J. G. Kirkwood[4], P. Debye[5], P. J. Flory[6], W. H. Stockmayer[7-10], H. A. Kramers[11], and B. H. Zimm[12, 13] who introduced rigorous mathematical arguments thus ensuring the change of polymer science from a mostly empirical to a quantitative field of research. Today we recognize with admiration that most of the principal mathematical implications have been fully disclosed by these authors.

This article deals with one of the above mentioned subjects already treated in the 1940's: branched polymers. We present a survey of a number of *scattering functions for special branched polymer structures*. The basis of these model calculations is still the *Flory-Stockmayer (FS) theory*[7, 14, 15] but now endowed with the more powerful technique of cascade theory which greatly simplifies the calculations.

The cascade theory is probably the oldest branching theory. It was developed by the English chaplain, the Reverend Watson[16, 18] and the biometrician Galton[17, 18] in 1873 who were evidently stimulated by Darwin's famous book on "The Origin of Species". Nowadays cascade theory is widely used in evolution theory[19, 20], in actuarial mathematics (birth and death processes), in the physics of cosmic ray showers and in the chemistry of combustion due to branched chain reactions[21-24].

In 1962 this mathematical technique was adapted to polymer science by M. Gordon[25] when he introduced a slight but essential alteration into this theory[26, 27] (inequivalence of the initial (zero-th) generation to all the other branching generations). In the outline given by Gordon, the cascade theory produces the same results for the molecular weight averages M_w and M_n, the prediction of the gel point, the mass fraction of extractable subcritically branched molecules in the gel, and for the molecular weight of this sol fraction as derived by the original FS theory. However, much more complicated models could now be treated without introducing new assumptions[28-33].

In 1970 a new stage was reached in this theory by introducing a special statistical weighting to the monomeric repeating units[34-36]. It turned out that this weighting produces a z-average over conformational properties. It is just this average that is measured by light-scattering techniques.

Two quantities are of particular interest:
1) the particle scattering factor of the static light- or neutron scattering $P_z(q)$
2) the first cumulant of the time correlation function (TCF) of the dynamic structure factor.

Knowing these functions, the mean-square radius of gyration $\langle S^2 \rangle_z$ and the translational diffusion coefficient D_z can easily be derived; eventually by application of the *Stokes-Einstein relationship* an effective hydrodynamic radius may be evaluated. These five quantities together with the two molecular weight averages M_w and M_n and the amount (weight fraction) of the sol provide an array of data which are most informative both for the architecture and molecular polydispersity of a branched system, as will be shown below.

II) *Branching is a widespread phenomenon in nature.* With delight we admire for instance the large variety of graceful branching in herbs and plants, and we may wonder which topologic rules govern such patterns. Peter S. Stevens[242], director of architectural planning office, followed this question more deeply in his beautiful book "Patterns in Nature". It is certainly disappointing that a similar direct approach is not possible with branched molecules in solution. This has two main reasons:

1. First, these objects from the microcosmos are so tiny that we have to take much care not to change involuntarily the molecular conformation by exerting external forces while we try to envisage the molecules. This, in particular, is true for the celebrated electron microscopy.

2. Second, even if we can make molecules visible in their natural conformation, we are looking at an ensemble of objects which have
   (i)   a large variation in size,
   (ii) a vast variety of different isomeric branched structures; additionally,
   (iii) each isomer can appear in many different conformations.

This immense variety is the result of the chemical conditions of synthesis which are mostly based on random reactions.

Experimentally, only *averages* over the ensemble and over time intervals can be observed. These averages are, however, not self-interpreting for branched molecules, and rules can be found only from the consideration of models:

In the first instance, a treatment of highly idealized models will be useful, which produces some rules of a certain universality.

In a more advanced stage, the models should be related to the actual chemical condition as closely as possible, which means that we have to give up the claim of universality when we turn to special problems.

This article shows how successfully the cascade branching theory works for systems of practical interest. It is a main feature of the Flory-Stockmayer and the cascade theory that all mentioned properties of the branched system are exhaustively described by the probabilities which describe how many links of defined type have been formed on some repeating unit. These *link probabilities* are very directly related to the extent of reaction which can be obtained either by titration (e.g. of the phenolic OH and the epoxide groups in epoxide resins based on bisphenol A[206, 207]), or from kinetic quantities (e.g. the chain transfer constant and monomer conversion[106, 107, 116]). The time dependence is fully included in these link probabilities and does not appear explicitly in the final equations for the measurable quantities.

Branching leads in many cases to *gelation and network formation.* Sometimes only precursors, i.e. synthetic resins, are wanted where gelation has to be prevented. Here, of course, a theory is most efficient which contains explicitly the chemical parameters responsible for the branching reactions, which can be altered in a similar manner as in real gelling reactions. This again is warranted by the close relation of the link probability to the extent of reaction (branching).

In other cases, the network structure is of greater interest; but networks are *solids*, in a way, where the number of analytic methods is limited. Mostly, the effects of rubber elasticity are measured by either mechanical or dielectric relaxation techniques. However, the relation between visco-elasticity and structure is complex and not fully explored and understood, in spite of the immense effort in the past decades. Much of the characteristic structure of the eventually formed network must already exist in the branched molecules in a system in the pre-gel state or in the sol fraction of the system in the post-gel state[179]. The number of experimental techniques used for the analysis of *soluble molecules* is much larger than for solids, and these techniques have in addition the advantage that only negligible external forces are exerted on the molecules. These conditions are almost ideally realized not only in static but also in dynamic light scattering measurements which are performed under conditions of chemical and physical equilibrium (in contrast to all other dynamic techniques where fairly large deviations from equilibrium are produced by the experimentalist and the return to equilibrium is measured).

Unfortunately, in light scattering we are not envisaging the objects themselves, i.e. the molecules, but observe a coded image: the *Fourier transform* of the system. The mathematics of Fourier transforms is well developed and offers no difficulties; but problems arise from the calculation of the required ensemble averages which are actually measured. Here again, the cascade theory provides us with a powerful method to derive these ensemble averages of the Fourier transforms. The labour which has to be invested for learning a new procedure is greatly rewarded by the facility of predicting properties from the chosen chemical starting conditions, which otherwise have to be determined empirically by extensive analytical research work. Optimization of a special reactions can now be better and more easily achieved by this theory than in former days.

III) *The Flory-Stockmayer and the equivalent cascade theory are not the only branching theories*, and a few words have to be said about the others.
1. For example, in recent years Macosko and Miller (MM)[37-40] have developed an attractively simple method which at first sight appears to be basically new. However, a closer inspection reveals the *MM approach* as being a degenerate case of the more general cascade theory. The simplicity is unfortunately gained at the expense of generality, and up-to-date conformation properties are not derivable by the MM-technique.
2. Another branching theory now frequently applied by physicists is the bond percolation on a lattice in space[41-45]. The *percolation theory* differs essentially from the FS theory both in its starting assumption and in the results deduced. Currently no full agreement could be reached on the justification of the basic assumptions in the two competing theories. Calculations of conformational properties and of the scattering functions are in principle possible but have not yet been carried out extensively. Physicists who apply the percolation theory have raised serious objections against the off-lattice FS theory which, incorrectly, is considered by these authors as being a typical representative of the so called *mean-field approximation*. They assume that the FS theory is not capable for various reasons of giving a reliable picture of the properties near the critical point of gelation. This point will be discussed in some detail later in this review. Here we only wish to point out that the FS theory has been widely misunderstood. In the 1940's Flory[14, 15] and Stockmayer[7] treated only the simplest

cases (comparable to the ideal gas treatment in statistical thermodynamics). These simple cases appear to be equivalent to a mean field approximation. Basically, however, a mean field approximation is not required in the FS theory, and this article will give some examples where correlations between neighbours are explicitly taken into account[29, 30].

The theory of branching is at present much more advanced than experiment. This fact is not surprising since a comprehensive characterization of a branched polymer involves appreciably more work than the corresponding characterization of a linear product. A few results are, however, already available. These experiments, so far, furnish on the one hand evidence for the validity of the FS theory and, on the other hand no convincing indication for a percolation theory on a lattice in space. This statement seems to hold even for branched polymers in good solvents where excluded volume effects may expected to exert a strong influence on the unperturbed dimensions which reflect the underlying *Gaussian chain statistics.* The reasons for the surprisingly low perturbation of the conformations is largely unknown and will have to be explored in future studies.

IV) A few words concerning the disposition of this review may be useful. Chapter B gives the basic relationships for static and dynamic light scattering and ends with the result that the mean-square radius of gyration $\langle S^2 \rangle_z$, the diffusion coefficient D_z, and the angular dependence of the first cumulant in the time correlation function Γ can be expressed in terms of the particle-scattering factor $P_z(q)$ if Gaussian statistics are assumed for the subchains connecting two monomeric units in the macromolecule.

The main purpose of this article is a comparison of branching theories with experimental results. Thus, Chap. C deals with the question how the unpleasant double summation, prescribed by the *basic light scattering (LS) theory,* can be handled and simplified. Graphical representations are helpful to overcome the abstractness of formulae, and use is made of this means as much as possible. In particular, the *"rooted tree"* will turn out as the most natural graph for a clear representation of branched structures and the underlying statistics which is efficiently covered by Gordon's branching theory. This chapter C presents the basis of the cascade theory, but the details are not absolutely needed for the understanding of the following chapters. A reader who is predominantly interested in the interpretation of data, may skip this chapter and turn immediately to Chap. D without losing too much of information.

Chapter D gives details on the common evaluation and interpretation of scattering experiments. Many experimental results are discussed in comparison with the behavior predicted by theory. This will show how much of this behavior can already be described by the cascade branching theory in spite of its obvious limitations. Furthermore, the great advantage of a combined measurement of both the static and dynamic LS is shown.

Finally, in the last Chap. E the more complex reactions are treated which are observed in *free-radical polymerization* and in *vulcanization of chains.* In the course of branching the experimentalist is often confronted with inhomogeneities in branching and chain flexibility and with chemical heterogeneity and steric hindrance due to an over-crowding of segments in space. Some of these problems of great practical importance have been solved in the past and are briefly reported.

B. Basic Equations for the Static and Dynamic Structure Factors

In this section some details of the static and dynamic structure factors and on the first cumulant of the time correlation function are given. The quoted equations are needed before the cascade theory can be applied. This section may be skipped on a first reading if the reader is concerned only with the application of the branching theory.

I. The Static Structure Factor and the Particle Scattering Factor[46-56]

1. Monodisperse Systems

Consider a molecular structure as shown for instance in Fig. 1. This polymer may be composed of x repeating units with dimensions that are small compared to the wave length of the incident primary beam, so that each unit can be considered as a point scatterer. Let $\mathbf{r}_j$ be the radius vector[1] of the j-th element from the origin. Then the scattered electric field of the x elements in the polymer is given by

$$E_s(q) = \sum_{j=1}^{x} E_{sj}(q) \propto \Delta\alpha \sum_{j=1}^{x} \exp(i\mathbf{q} \cdot \mathbf{r}_j) \tag{B.1}$$

and the corresponding scattering intensity is

$$i_x(q) = \langle |E_s(q) E_s^*(q)| \rangle = \langle E(0) E^*(0) \rangle S(q)/x^2 \tag{B.2}$$

where $S(q)$ is the static structure factor of an isolated molecule which according to Eq. (B.1) is defined as

$$S(q) = \sum_{j=1}^{x} \sum_{k=1}^{x} \langle \exp(i\mathbf{q} \cdot \mathbf{r}_{jk}) \rangle \tag{B.3}$$

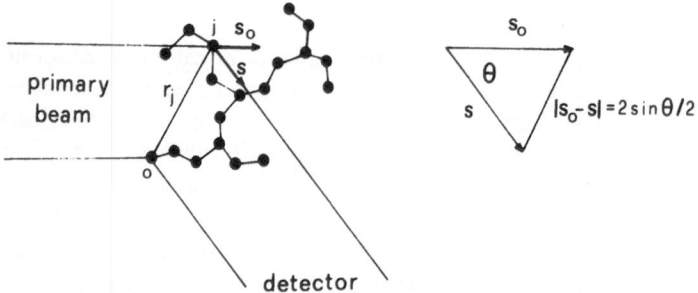

Fig. 1. Scattering of light from a branched particle with dimensions greater than $\lambda/20$. S_0 and S are unit vectors in direction of the primary beam and the scattered light, respectively; the phase difference of the two scattered waves emerging from element j and o is given by $\mathbf{q} \cdot \mathbf{r}_j = \mathbf{k} \cdot \mathbf{r}_j \cdot |S_0 - S| = (4\,\pi/\lambda) \sin \theta/2$, where Θ is the scattering angle, and $k = 2\,\pi/\lambda$

---

1 Bold characters and numbers are vectors or matrices and tensors, respectively

with $r_{jk} = r_j - r_k$. In these equations q is the scattering vector which is defined by the directions of the incident scattered rays s_0 and s respectively as

$$q = (2\pi/\lambda)(s_0 - s) \tag{B.4}$$

Its magnitude is (see Fig. 1, right hand side)

$$q = (4\pi/\lambda)\sin\theta/2 \tag{B.5}$$

where θ is the scattering angle. The angle brackets denote the ensemble average over all orientations and distance fluctuations. Finally $\Delta\alpha = \alpha - \langle\alpha\rangle$ describes the deviation of the polarizability from its equilibrium value. The normalized structure factor is called the *particle-scattering factor*[57, 58]

$$P_x(q) \equiv S(q)/S(0) = S(q)/x^2 \tag{B.6}$$

Equation (B.2) is valid for one isolated molecule in solution. For dilute solutions the fluctuation theory shows that[59]

$$\langle\Delta\alpha^2\rangle \propto (n_s dn/dc)^2 RTc/(\partial\pi/\partial c) \tag{B.7}$$

Thus, for very dilute solutions where no specific phase relationships exist, one has

$$i(q) \propto cM(n_s dn/dc)^2 P_x(q) \tag{B.8}$$

Introducing the Rayleigh ratio of the scattering intensity to the primary beam intensity $R(q) \equiv i(q)r^2/I$, where r is the distance of the detector from the centre of the scattering cell, Eq. (B.8) may also be written as

$$R(q) = KcMP_x(q) = KcMx^{-2}\sum_j^x\sum_k^x\langle\exp(iq\cdot r_{jk})\rangle \tag{B.9}$$

with a constant K that describes the "contrast" of the scattering intensity of the solute over that of the pure solvent. When vertically polarized incident light is used K is given as[2]

$$K = (4\pi^2/(\lambda_0^4 N_A))(n_s dn/dc)^2 \tag{B.10}$$

In these equations n_s is the solvent refractive index, dn/dc the refractive index increment, c the polymer concentration in g/ml, T the temperature in K, R the gas constant, N_A Avogadro's number, and π the osmotic pressure. Equation (B.8) follows from Eq. (B.7) by using the familiar virial expansion of the osmotic pressure

$$\pi = RTc(1/M + A_2c + A_3c^3 + ...) \tag{B.11}$$

---

2 When using horizontally polarized light, Eq. (B.10) is multiplied by the polarization factor $\cos^2\theta$, while with the use of unpolarized light the polarization factor is $(1 + \cos^2\theta)/2$

For copolymers, each scattering element j scatters with an amplitude of Δa_j. We now assume that Δa_j is proportional to the molecular weight of the individual monomeric units M_{0j} and proportional to the corresponding refractive index increment $v_j = (dn/dc)_j$. Instead of Eq. (B.9), the scattering intensity is now given by[60, 61]

$$R(q) = Kc/(v^2 M) \sum_{j}^{x} \sum_{k}^{x} \langle \exp(i\mathbf{q} \cdot \mathbf{r}_{jk}) \rangle_{\iota} M_{0j} M_{0k} v_j v_k \qquad (B.12)$$

and

$$P(q) = R(q)/R(0) \qquad (B.13)$$

where the total refractive index increment v of the copolymer is related to the v_j of the different components as

$$v = w_A v_A + w_B v_B + \dots \qquad (B.14)$$

$$w_A + w_B + \dots = 1$$

with $w_A, w_B, \dots$ the weight fractions of the components A, B etc.

2. Polydisperse Systems

In general, a solution contains molecules of different molecular weights, and in this case the scattering intensity is the sum of the contributions from the various molecules of molecular weight M_x with the concentration c_x or weight fraction $w_x = c_x/c$. The total concentration is $c = \sum c_x$. For homopolymers, one now obtains

$$R(q) = Kc \sum_{x=1}^{\infty} w_x M_x \left[x^{-2} \sum_{j}^{x} \sum_{k}^{x} \langle \exp(i\mathbf{q} \cdot \mathbf{r}_{jk}) \rangle \right] \qquad (B.15)$$

or

$$R(q) = KcM_w P_z(q) \qquad (B.16)$$

where

$$M_w = \sum w_x M_x \qquad (B.17)$$

$$P_z(q) = \left(\sum w_x M_x P_x(q) \right) / \left(\sum w_x M_x \right) \qquad (B.18)$$

M_w is the weight average molecular weight and $P_z(q)$ the z-average of the particle-scattering factor; the particle-scattering factor of an x-mer in the ensemble is given by the expression in the brackets of Eq. (B.15).

For copolymers, the corresponding equation for the scattering intensity reads[60, 61]

$$R(q) = Kc \sum_{\text{all } xi} w_{xi}/(v^2 M_{xi}) \left[\sum_{j}^{x} \sum_{k}^{x} \langle \exp(i\mathbf{q} \cdot \mathbf{r}_{jk}) \rangle M_{0j} M_{0k} v_j v_k \right]_i \qquad (B.19)$$

Here w_{xi} denotes the weight fraction of an x-mer with a special composition and v is given by Eq. (B.14). The relationship Eq. (B.19) may conveniently be written as

$$R(q) = KcM_w^{app}P_z(q) = KcM_wP_z^{app}(q) \tag{B.20}$$

where

$$M_w = \sum w_{xi}M_{xi} \tag{B.21 a}$$

$$M_w^{app} = \sum (w_{xi}/M_{xi} v^2) \sum_j \sum_k M_{0j}M_{0k} v_j v_k = \sum (w_{xi}M_{xi} v_i^2/v^2) \tag{B.21 b}$$

$$P_z(q) = \frac{\sum w_{xi}/M_{xi})[\sum\sum \langle \exp(i\mathbf{q} \cdot \mathbf{r}_{jk})\rangle M_{0j}M_{0k} v_j v_k]_i}{\sum w_{xi}/M_{xi})[\sum\sum M_{0j}M_{0k} v_j v_k]_i} \tag{B.21 c}$$

$$P_z^{app}(q) = \frac{\sum w_{xi}/M_{xi} v^2)[\sum\sum \langle \exp(i\mathbf{q} \cdot \mathbf{r}_{jk})\rangle M_{0j}M_{0k} v_j v_k]_i}{\sum w_{xi}M_{xi}} \tag{B.21 d}$$

In the second part of Eq. (B.21 b), the refractive index increment of the i-th isomer has been used which is defined in a way similar to Eq. (B.14), i.e.

$$v_i = w_{Ai}v_A + w_{Bi}v_B + \dots \tag{B.14'}$$

where the w_{Ai}, w_{Bi} etc. are the weight fractions of the components in the special isomer.

The Eqs. (B.15) and (B.19) are the basic relationships for the calculation of static conformational properties. Having solved the various sums occuring in these relationships, the correct and apparent molecular weight is obtained by setting q = 0, and the mean-square radii of gyration are the corresponding first coefficients in the series expansion of $M_wP_z(q)$ and $M_wP_z^{app}(q)$, respectively

$$M_w \langle S^2 \rangle_z = -3 d [M_wP_z(q)]/dq^2|_{at\ q=0} \tag{B.22 a}$$

and

$$M_w \langle S^2 \rangle_z^{app} = -3 d [M_wP_z^{app}(q)]/dq^2|_{at\ q=0} \tag{B.22 b}$$

In most cases we will assume Gaussian statistics for the subchains connecting two units j and k in the molecule. These two points have then an end-to-end distribution of

$$W(\mathbf{r}_{jk}) = (3/2\pi^2 \langle r_{ij}^2 \rangle)^{3/2} \exp(-3 r_{jk}^2/2 \langle r_{jk}^2 \rangle) \tag{B.23}$$

With this distribution, the average in Eq. (B.15) or (B.3) becomes[62]

$$\langle \exp(i\mathbf{q} \cdot \mathbf{r}_{jk})\rangle = \exp(-b^2q^2n/6) \tag{B.24 a}$$

when n is the number of monomer units in the chain connecting element j with element k, and b^2 is the effective bond length. For copolymers, we have to specify the number n of units with bond length b_A and number m of units with bond length b_B etc. which form the length of the subchain. For a binary copolymer, we have

$$\langle \exp(iq \cdot r_{jk}) \rangle = \exp[-(b_A^2 n + b_B^2 m) q^2/6] \equiv \phi_{jk}^{IS} \tag{B.24 b}$$

where ϕ_{jk}^{IS} is used as an abbreviation for the contribution of a pair of scattering elements j, k to the static or frequency integrated light scattering (IS). Inserting Eq. (B.24 b) into Eq. (B.21 c) one has

$$P_z(q) = \frac{\sum (w_{xi} M_{xi}) [\sum \sum \exp(-yq^2) M_{0j} M_{0k} v_j v_k]_i}{\sum (w_{xi}/M_{xi}) [\sum \sum M_{0j} N_{0k} v_j v_k]_i} \tag{B.21'}$$

with

$$y = (b_A^2 n + b_B^2 m)/6$$

II. The Dynamic Structure Factor and its First Cumulant[63-69]

1. The Time Correlation Function

In dynamic or quasi-elastic light scattering, a time dependent correlation function $\langle i(0) i(t) \rangle \equiv G_2(t)$ is measured, where i(0) is the scattering intensity at the beginning of the experiment, and i(t) that at a certain time later. Under the conditions of dilute solution (independent fluctuation of different small volume elements), the intensity correlation function can be expressed in terms of the electric field correlation function $g_1(t)$

$$g_1(t) = \frac{\langle |E_s^*(0) E_s(t)| \rangle}{\langle |E_s^*(0) E_s(0)| \rangle} = S(q, t)/S(q) \tag{B.25}$$

as follows

$$G_2(t) = A + B g_1^2(t) \tag{B.26}$$

where A and B are constants. Figure 2 gives an example for the time correlation function (TCF) $G_2(t)$ and for $\ln g_1(t)$.

The denominator in Eq. (B.25) is recognized as the static scattering function, and the numerator is called the dynamic structure factor $S(q, t)$ which for a homopolymer is given as

$$S(q, t) = \sum_j^x \sum_k^x \langle \exp(iq \cdot r_{jk}(t)) \rangle \tag{B.27}$$

with

$$r_{jk}(t) = r_j(0) - r_k(t) \tag{B.28}$$

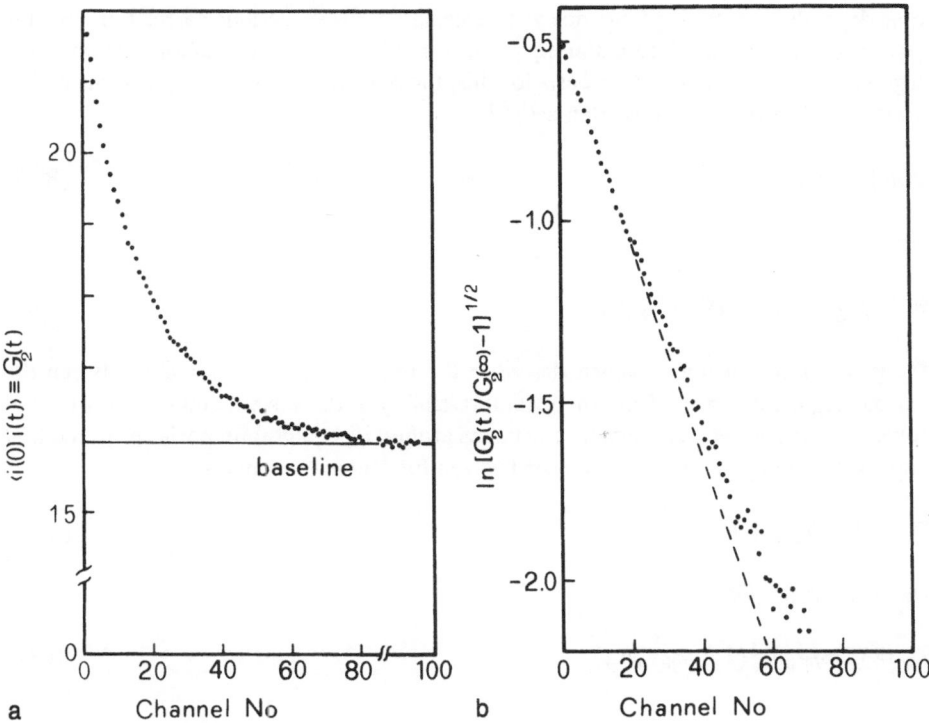

Fig. 2. a. Intensity correlation of the scattered light from a PMMA sample in acetone recorded with a Malvern autocorrelator with 96 channels. At channel 80, a time shift device is introduced, by which the last 12 channels are shifted by 256 times the delay time of the first channel. These last channels were used for a determination of the base line. **b** Plot of ln $(G_2 - 1)^{1/2}$ against the channel number (= time) for the same sample as in **a**. The deviation from the straight line arises from the internal flexibility of the PMMA chain

The angle brackets now denote the average over the space-time distribution, i.e. the average has to be taken over all possible positions of the j-th element at time zero and over all possible positions of the k-th element at a delayed time t.

For a polydisperse system of homopolymers, one finds in the same manner as outlined for the static scattering[66, 70−72]

$$g_1(t) = \sum_{x}^{\infty} w_x M_x P_x(q) g_{1x}(t) / (\sum w_x M_x P_x(q)) \tag{B.29}$$

and for copolymeric systems, one has with Eq. (B.19)

$$g_1(t) = \frac{\sum (w_{xi}/M_{xi})[\sum\sum \langle \exp(iq \cdot r_{jk}(t)) \rangle M_{0j} M_{0k} v_j v_k]_i}{\sum (w_{xi}/M_{xi})[\sum\sum \langle \exp(iq \cdot r_{jk}(0)) \rangle M_{0j} M_{0k} v_j v_k]_i} \tag{B.30}$$

We have already given in Eq. (B.24) the average of the exponential function for the static scattering assuming Gaussian statistics. Before the corresponding average for the time dependent scattering function can be calculated, the space-time correlation function $\psi(r, t)$ or at least the differential equation of motion for the macromolecule and their segments must be known. As a basis for this, the *Kirkwood diffusion equation* is mostly used[73]. The latter may be written as[74, 75]

$$\partial\psi/\partial t = \mathcal{D}\psi \tag{B.31}$$

with

$$\mathcal{D} = \sum\sum \nabla_j \cdot \mathbf{D}_{jk} \cdot [\nabla_k - \nabla_k \psi_0] \tag{B.32}^3$$

$\mathbf{D}_{jk}$ is the diffusion tensor which describes the hydrodynamic interaction between the various segments, arising from the incompressibility of the solvent and the back flow of solvent molecules when a segment fluctuates around its equilibrium position. In the first approximation by Oseen[76] this tensor is given for a homopolymer as

$$\mathbf{D}_{jk} = kT[\delta_{jk}/\zeta + (1 - \delta_{jk})\mathbf{T}_{jk}] \tag{B.33}$$

where $\mathbf{T}_{jk}$ is the Oseen tensor

$$\mathbf{T}_{jk} = (8\pi\eta_0 r_{jk})^{-1}(\mathbf{1} + \mathbf{r}_{jk}\mathbf{r}_{jk}/r_{jk}^2) \tag{B.34}$$

2. Cumulant Expansion

An explicit solution of Eq. (B.31) is possible for macromolecular rings which obey Gaussian statistics[75]. For open linear and branched molecules, only approximate solutions are known so far[69]. One of these approximations is the so called *cumulant expansion* of $S(q, t)$[77, 78], which is a series expansion of the logarithmic TCF in powers of the delay time t

$$\ln(S(q, t)) = \ln(S(q)) - \Gamma_1 t + \Gamma_2 t^2/2 - \ldots + (-1)^n \Gamma_n t^n/n! \tag{B.35}$$

As was shown by Bixon[75], Ackerson[79] and Akcasu and Gurol[80], the first cumulant can be calculated exactly without knowing the space-time distribution function with the following result

$$\Gamma_1 = -\partial\ln(S(q, t)/\partial t)_{t=0} = -\partial\ln(g_1(t)/\partial t = \frac{\sum\sum \langle(q \cdot \mathbf{D}_{jk} \cdot q)\exp(i q r_{jk})\rangle}{\sum\sum \langle\exp(i q r_{jk})\rangle} \tag{B.36}$$

This equation holds for monodisperse homopolymers; for polydisperse copolymers one has

$$\Gamma_1 = \frac{\sum (w_{xi}/M_{xi})[\sum\sum \langle q \cdot \mathbf{D}_{jk} \cdot q)\exp(i q \cdot r_{jk})\rangle M_{0j}M_{0k} v_j v_k]_i}{\sum (w_{xi}/M_{xi})[\sum\sum \langle\exp(i q \cdot r_{jk})\rangle M_{0j}M_{0k} v_j v_k]_i} \tag{B.37}$$

---

3 Grotesque character denotes an operator

Note, that non only the ensemble average is needed. In the following we confine ourse-lves to the first cumulant, i.e. to the initial part of the TCF, and drop the subscript 1 in Eq. (B.37).

3. The First Cumulant

For Gaussian subchains, the average $\langle (\mathbf{q} \cdot \mathbf{D}_{jk} \cdot \mathbf{q}) \exp(i\mathbf{q} \cdot \mathbf{r}_{jk}) \rangle$ can be solved exactly, and the result will be given later in this section. However, this exact solution is already rather complicated even for this single pair of scattering elements and makes the summa-tion over all pairs of scattering element for complex molecular structures formidable. It is tempting, therefore, to use a pre-average approximation in which the correct average is replaced by

$$q^{-2}\phi_{jk}^Q \equiv \langle (\mathbf{q} \cdot \mathbf{D}_{jk} \cdot \mathbf{q}) \exp(i\mathbf{q} \cdot \mathbf{r}_{jk}) \rangle \simeq \langle \mathbf{q} \cdot \mathbf{D}_{jk} \cdot \mathbf{q} \rangle \langle \exp(i\mathbf{q} \cdot \mathbf{r}_{jk}) \rangle = \phi_{jk, \text{pre}}^Q$$

The superscripts Q stands here for quasi-elastic scattering, and the subscript "pre" means the pre-average approximation.

Before giving the explicit equations for the various averages, it will be useful to consider the limit of $q \to 0$. Theory on dynamic scattering proves that in this limit

$$\lim_{q \to 0} \Gamma = Dq^2 \qquad (B.38)$$

where D is the translational diffusion coefficient. For a polydisperse copolymer, this diffusion coefficient is the z-average of an apparent diffusion coefficient which reduces to the true D_z only for polymers of identical composition. Making use of this limiting behavior, one obtains from Eq. (B.36).

$$D = q^{-2} \sum_j^x \sum_k^x \langle \mathbf{q} \cdot \mathbf{D}_{jk} \cdot \mathbf{q} \rangle / x^2 \qquad (B.39)$$

which is exactly Kirkwood's general diffusion equation[81] derived from the *transport equation* only by applying hydrodynamic pre-averaging. We recognize now that from the theory of dynamic light scattering, the result is independent of a pre-average approxima-tion; however, D in Eq. (B.39) is a diffusion coefficient at zero time (by definition of the first cumulant) and is not necessarily the same as that obtained from macroscopic trans-port processes.

We now give the formulae for the various averages. For rigid particles the average has to be performed only over all orientations (subscript "or"). This yields

$$\langle \exp(i q r_j) \rangle_{or} = \sin z / z \qquad (B.40)$$

$$\langle \mathbf{q} \cdot \mathbf{D}_{jk} \cdot \mathbf{q} \rangle_{or} = q^2 kT [\delta_{jk}/\zeta_j + (1 - \delta_{jk})/(6\pi\eta_0 r_{jk})] \qquad (B.41)$$

$$\langle (\mathbf{q} \cdot \mathbf{D}_{jk} \cdot \mathbf{q}) \exp(i\mathbf{q} \cdot \mathbf{r}_{jk}) \rangle_{or}$$
$$= q^2 kT (\delta_{jk}/\zeta_j + (1 - \delta_{jk})/(4\pi\eta_0 r_{jk}) \times (\sin(z)/z + \cos(z)/z^2 - \sin(z)/z^3) \qquad (B.42)$$

with $z = \mathbf{q} \cdot \mathbf{r}_{jk}$.

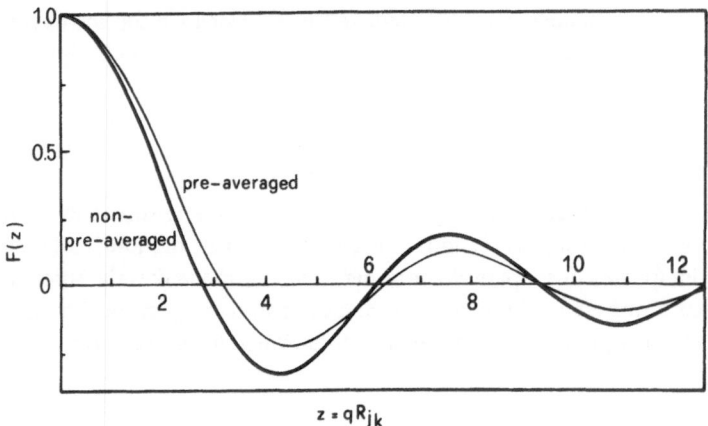

Fig. 3. The dynamic scattering function for a pair of elements (j, k) after averaging over all orientations in the pre-average approximation of the hydrodynamic interaction, and without this approximation

Figure 3 demonstrates the effect of the pre-averaging for the rigid structure where

$F(z) = \sin(z)/z$ for the pre-average

$F(z) = (3/2)\sin(z)/z - \cos(z)/z^2 + \sin(z)/z^3)$ for the non pre-average

For flexible structures, the function $F(z)$ has to be averaged over the distance distribution. For Gaussian chains one obtains[82]

$$\langle \exp(i\mathbf{q} \cdot \mathbf{r}_{jk}) \rangle = \exp(-Yq^2) \equiv \phi_{jk}^{IS} \tag{B.43}$$

$$q^{-2} \langle \mathbf{q} \cdot \mathbf{D}_{jk} \cdot \mathbf{q} \rangle = (kT/\zeta_j)\delta_{jk} + (1 - \delta_{jk})AqY^{-1/2} \equiv \phi_{jk}^{D} \tag{B.44}$$

$$q^{-2} \langle (\mathbf{q} \cdot \mathbf{D}_{jk} \cdot \mathbf{q})\exp(i\mathbf{q} \cdot \mathbf{r}_{jk}) \rangle = (kT/\zeta_j)\delta_{jk} + (1 - \delta_{jk})AY^{-1/2}(3/4)H(Yq^2) \equiv \phi_{jk}^{Q} \tag{B.45}$$

where, for the example of a binary copolymer,

$$Y = (nb_A^2 + mb_B^2)/6 \tag{B.46}$$

$$A = kT/(6\pi^{3/2}\eta_0) \tag{B.46}$$

$$H(Yq^2) = [2/Yq^2 + (Yq^2)^{-3}]D(Yq^2) - (Yq^2)^{-2} \tag{B.47}$$

$$D(Yq^2) = \exp(-Yq^2)\int_0^{Yq^2}\exp(t^2)\,dt \tag{B.48}$$

n and m are the numbers of the monomer units of the one and the other component in the subchain connecting element j with k.

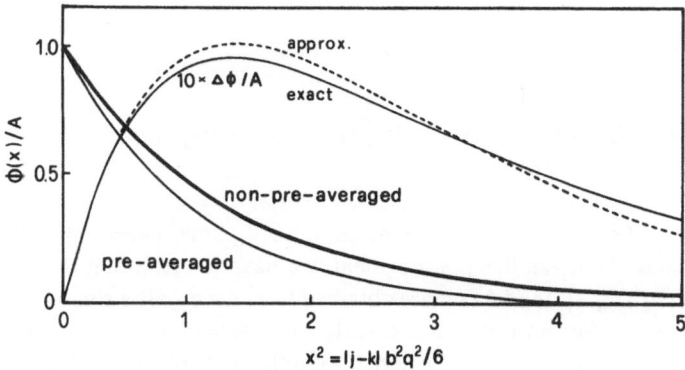

Fig. 4. The dynamic scattering function ϕ_{jk}^Q for a pair of elements (j, k) after averaging over all orientations and distance fluctuations in the hydrodynamic pre-average approximation, Eq. (B.49), and without this approximation, Eq. (B.45). The line labeled "exact" gives the exact deviation of the preaverage approximation, the dotted line represents the approximation of Eq. (B.50)[82]

For the pre-average approximation, we have simply[80, 82] (compare Eq. (B.45))

$$\phi_{jk, \, pre}^Q = \phi_{jk}^D \, \phi_{jk}^{IS} = [(kT/\zeta_j) \, \delta_{jk} + (1 - \delta_{jk}) \, AqY^{-1/2}] \exp(-Y \, q^2) \tag{B.49}$$

In Fig. 4, the function $\phi(Y)/A$ is plotted against Yq^2 for the correct quasielastic pair function and for the pre-average approximation, where the free draining term $(kT/\zeta_j) \, \delta_{jk}$ has been neglected. The deviations between the two curves can be approximated by[82] (see Fig. 4).

$$\Delta \phi_{jk}^Q = \phi_{jk}^Q - \phi_{jk}^D \, \phi_{jk}^{IS} = (1/5) \, AYq^2 \exp(-0.72 \, Yq^2) \tag{B.50}$$

4. Integral Representation of the First Cumulant

Neglecting the free draining part, the Eqs. (B.45), (B.49) and (B.50) may be written in the following equivalent integral form[83, 84]

$$\phi_{jk}^D \quad = (2 \, A/\pi^{1/2}) \int_0^\infty \exp(-Yq^2) \, dq \tag{B.51}$$

$$\phi_{jk, \, pre}^Q = (2 \, A/\pi^{1/2}) \int_0^\infty \exp(-Yq^2 - Y\beta^2) \, d\beta \tag{B.52}$$

$$\Delta \phi_{jk}^Q \quad = (A/5 \, \pi^{1/2}) \, q^2 \int_0^\infty \frac{d}{\beta^2} \, [\exp(-0.72 \, Yq^2) - \exp(-0.72 \, Yq^2 - \beta^2)] \tag{B.53}$$

where use was made of the identities

$$Y^{-1/2} = (2/\pi^{1/2}) \int_0^\infty \exp(-Yq^2)\, dq \tag{B.54}$$

$$Y^{1/2} = -0.5 \int_0^y t^{-1/2}\, dt = \pi^{-1/2} \int_0^\infty (1 - \exp(-Y\beta^2))\, d\beta/\beta^2 \tag{B.55}$$

These integral representations of the ϕ_{jk} have a great advantage in numerical calculations. To make this point evident, we have to recall that the measurable quantities D_z, $(\Gamma/q^2)_{\text{preaverage}}$ and $\Delta\Gamma/q^2$ are obtained from the ϕ_{jk} after summation over all pairs (see next chapter for more details). Clearly, the evaluation of a geometric series is more easily performed than that of a series with elements of the kind $|j-k|^{-1/2}$ or $|j-k|^{1/2}$. With the identities of Eq. (B.51) to (B.53), we obtain for D_z and the first cumulant integrals over a geometric series which essentially is the double sum of the particle scattering factor. Hence, when the relationships (B.51) to (B.53) are inserted in the formulae for $D_{z,\text{app}}$ (Eq. (B.39)) and the first cumulant (Eq. (B.37)), we arrive at the very useful equations

$$D_{z,\text{app}} = (2\,A/\pi^{1/2}) \int_0^\infty P_z(Yq^2)\, dq \tag{B.56}$$

$$\frac{\Gamma}{q^2} = (A/P_z(Yq^2)\,\pi^{1/2})\left\{ 2 \int_0^\infty P_z(Yq^2 + Y\beta^2)\, d\beta - 0.2\,q^2 \int_0^\infty \beta^{-2}[P_z(0.72\,Yq^2) \right.$$

$$\left. - P_z(0.72\,Yq^2 + Y\beta^2)]\, d\beta \right\} \tag{B.56}$$

or

$$\Gamma/\Gamma_0 = \left[2\,P_z(Yq^2) \int_0^\infty P_z(Y\beta^2)\, d\beta \right]^{-1}$$

$$\times \left[2 \int_0^\infty P_z(Yq^2 + Y\beta^2)\, d\beta - q^2 0.2 \int_0^\infty \beta^{-2}[P_z(0.72\,Yq^2) - P(0.72\,Yq^2 + Y\beta^2)]\, d\beta \right] \tag{B.58}$$

where Γ_0 is the first cumulant at $q = 0$, and the expression $P_z(Yq^2 + Y\beta^2)$ means that the argument Yq^2 in the particle scattering factor has to be replaced by the argument $Y(q^2 + \beta^2)$.

Hence the problems in static and dynamic scattering reduce to the evaluation of the sums for the particle-scattering factor. All other quantities can be derived by differentiation, which yields the mean-square radius of gyration, or by integration, which yields the diffusion coefficient or the angular dependence of the first cumulant. However, the integral representation remains valid only when the Gaussian distance distribution for the subchains is obeyed[85, 86].

C. Evaluation of Double Sums

I. General Remarks

In the last chapter, equations were derived for the particle-scattering factor, the mean-square radius of gyration, the diffusion coefficient and the first cumulant of the dynamic structure factor. All these have the common feature that, for homopolymers at least, they can be written in the following form:

$$P_w \langle \phi \rangle_z = \sum_{x=1}^{\infty} w_x x \left[x^{-2} \sum_j^x \sum_k^x \phi_{jk} \right] \tag{C.1}$$

The equations for copolymers are a little more complicated but can be reduced to similar expressions, as will be shown later in this chapter. Moreover, if Gaussian statistics is obeyed for the subchains connecting two chain elements j and k, we have

$$\phi_{jk} = \exp \left(- (b^2 q^2 / 6) |j - k| \right) \equiv \phi^{|j - k|} \tag{C.2}$$

and in this case all conformational quantities can be derived from the particle-scattering factor $P(q^2)$.

Particle scattering factor, z-average:

$$P_w P_z (q^2) = \sum_{x=1}^{\infty} w_x x \left[x^{-2} \sum_j^x \sum_k^x \phi^{|j - k|} \right] \tag{C.3}$$

Degree of polymerization, weight average:

$$P_w = \lim_{q \to 0} (P_w P_z (q^2)) \tag{C.4}$$

Mean-square radius of gyration, z-average:

$$\langle S^2 \rangle_z = -3 (dP_z (q^2)/dq^2)|_{at \ q^2=0} \tag{C.5}$$

Translational diffusion coefficient, z-average:

$$D_z = (2 A / \pi^{1/2}) \int_0^{\infty} P(q^2) dq \tag{C.6}$$

with

$$A = kT/6 \pi^{3/2} \eta_0)$$

where k is Boltzmann's constant, T the temperature in K, and η_0 the solvent viscosity.

First cumulant, z-average:

$$\frac{\Gamma}{q^2} = (A/\pi^{1/2}P_z(q^2)\left\{2\int_0^\infty P_z(q^2 + \beta^2)\,d\beta - 0.2q^2\int_0^\infty \beta^{-2}[P_z(0.72\,q^2)-P_z(0.72\,q^2 + \beta^2)]\,d\beta\right\}$$

(C.7)

or

$$\Gamma/\Gamma_0 = \left[2P_z(q^2)\int_0^\infty P_z(\beta^2)\,d\beta\right]^{-1}$$

$$\times\left\{2\int_0^\infty P_z(q^2 + \beta^2)\,d\beta - 0.2q^2\int_0^\infty \beta^{-2}[P_z(0.72\,q^2) - P_z(0.72\,q^2 + \beta^2)]\,d\beta\right\}$$

(C.8)

Thus, the main problem is how the triple sum in Eq. (C.3) can be evaluated. At first sight, this problem looks formidable. In the following, the techniques of evaluation are described in some detail, starting with the simplest cases of monodisperse homopolymers and proceeding step by step to the more complex molecules of branched copolymers, which are highly polydisperse in molecular weight and heterogeneous in composition.

II. Monodisperse Homopolymers

1. Linear Chains; Debye's Technique[5, 87)]

For monodisperse homopolymers, Eq. (C.3) reduces to

$$x^2P(q^2) = \sum_j^x \sum_k^x \phi_{jk}$$

(C.9)

As long as loops are neglected, the double index in Eq. (C.9) always means the number of units in a subchain or a path that leads from element j to element k. Setting $n = j - k$, the double sum may be written in a squared area as follows:

$$S(q^2) = \begin{bmatrix} \phi_0 & +\phi_1 & +\phi_2 & +\ldots+\phi_n & +\ldots+\phi_{x-1} \\ +\phi_1 & +\phi_0 & +\phi_1 & +\ldots+\phi_{n-1}+\ldots+\phi_{x-2} \\ + \ldots & \ldots & \ldots & \ldots\ldots \ldots\ldots \\ +\phi_n & +\phi_{n-1}+\phi_{n-2}+\ldots+\phi_0 & +\ldots+\phi_{x-n} \\ + \ldots & \ldots & \ldots & \ldots\ldots \ldots\ldots \\ +\phi_{x-1}+\phi_{x-2}+\phi_{x-3}+\ldots+\phi_{x-n-1}\ldots+\phi_0 \end{bmatrix} = x^2P(q^2)$$

(C.10)

which yields

$$x^2P(q^2) = x\phi_0 + 2\sum_{n=1}^{x-1}(x-n)\phi_n$$

(C.11)

In particular for Gaussian chains, where $\phi_n = \phi^n$ and when $x \gg 1$

$$x^2 P(q^2) = x + [2\phi/(1 - \phi)] [x + \phi/(1 - \phi)(\phi^x - 1)] \qquad (C.12)$$

with

$$\phi = \exp(-b^2 q^2/6)$$

For $b^2 q^2/6 \ll 1$, which is the region of light scattering and of small-angle neutron or X-ray scattering, Eq. (C.12) becomes

$$P(q^2) = 2u^{-4} [(u^2 - 1) + \exp(- u^2)] \qquad (C.13)$$

which is Debye's famous equations for the particle scattering factor of linear randomly coiled chains, where

$$u^2 = xb^2 q^2/6$$

2. Rooted Tree Treatment; Regular Star Molecules[88]

Debye's technique is not always applicable and fails completely for complex, branched structures ("trees"). In such cases, a slightly different technique can be used. One special unit, say the j-th unit, can be selected as a reference point, and all pairs of units of the same path length can be grouped together, and the result may be summed over all path length occurring. We shall call the reference unit the "root of the tree". To get the total double sum, each unit has to be selected as a root and the result of the summation over the different trees has to be added. This leads to the final result

$$S(q^2) = \sum_j^x \sum_k^x \phi_{jk} = \left(1 + \sum_{j=1}^{x} \sum_{n=1}^{n_{max}} N_j(n) \phi_n\right) \qquad (C.14)$$

where $N_j(n)$ is the number of paths of length n for the j-th *such* tree. This *rooted tree treatment* may be illustrated by the example of regular stars. Figure 5 shows the two essentially different types of rooted trees, i.e. that with the branching unit as root and that where one of the units of a ray furnishes the root.

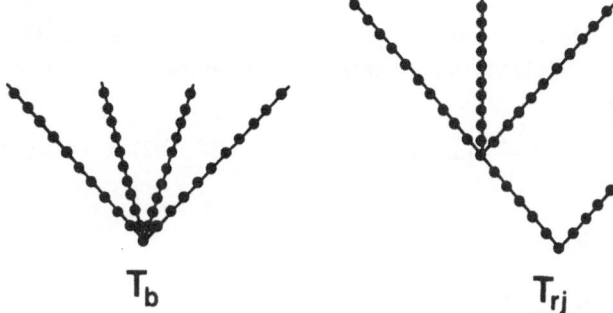

Fig. 5. Two typical rooted tree representations of a four ray star-molecule. T_b: the branch point selected as root; T_{rj}: the j-th element of a ray selected as root

For the tree with the branching unit as root, the number of paths is $N(n) = f$ for all path lengths to the root, thus

$$T_b = 1 + f \sum_{n=1}^{m} \phi_n \qquad (C.15)$$

For the trees where the j-th unit of a ray is chosen as a root, we have paths of multiplicity $N(n) = 1$ with (i) a maximum length of $m - j$ (the right branch in Fig. 5 emerging from the root) and (ii) a maximum length of j elements (left branch, including the branching point; finally, we have $N(n) = (f - 1)$ paths of the length $j + n$ with a maximum path length of $n_{max} = m$. Hence

$$T_{rj} = 1 + \sum_{n=1}^{m-j} \phi_n + \sum_{n=1}^{j} \phi_n + (f-1) \sum_{n=1}^{m} \phi_{n+j} \qquad (C.16)$$

The structure factor is the sum of all trees

$$S(q^2) = T_b + f \sum_{j=1}^{m} T_{rj} = T_b + fT_r \qquad (C.17)$$

with T_b given in Eq. (C.15) and

$$T_r = m + 2 \sum_{j}^{m-1} \sum_{n}^{j-1} \phi_n + \sum_{j}^{m} \phi_j + (f-1) \sum_{n}^{m} \sum_{j}^{m} \phi_{n+j} \qquad (C.18)$$

Assuming Gaussian statistics, we have $\phi_{n+j} = \phi^n \phi^j$ and thus

$$T_b = 1 + fP_1 \qquad (C.19\,a)$$

$$T_r = P_2 + P_1 + (f - 1) P_1^2 \qquad (C.19\,b)$$

with

$$P_1 = \phi(1 - \phi^m)/(1 - \phi) \qquad (C.19\,c)$$

$$P_2 = m + (2/(1 - \phi)) [(m - 1) - P_1] \qquad (C.19\,d)$$

For long rays, one has $\phi \cong 1, 1 - \phi \cong b^2 q^2/6$, and T_b and P_1 can be neglected compared to P_2. This leads to the following equation for the particle-scattering factor of regular stars

$$P(q^2) = \frac{2}{fu_r^4} [u_r^2 - (1 - \exp(-u_r^2)) + ((f - 1)/2)(1 - \exp(-u_r^2))^2] \qquad (C.20)$$

with

$$u_r^2 = mb^2q^2/6$$

Equation (C.20) was first derived by Benoit[89] and it reduces to the Debye equation of a linear chain for $f = 1$ and $f = 2$[90].

For the mean-square radius of gyration, one obtains after application of the differential operation (outlined in Eq. C.5)

$$\langle S^2 \rangle = (3f - 2)/(6f^2)\, b^2 m \tag{C.21}$$

and for the diffusion coefficient after performing the required integration (indicated in Eq. C.6)

$$D = (8/3\,f)\,(1 + (2^{1/2} - 1)\,(f - 1))\, A/(b m^{1/2}) \tag{C.22}$$

a result which was first derived by Stockmayer and Fixman[91].

By the same technique other regular structures can be calculated, and results are known for regular comb molecules[92], regularly branched trifunctional molecules[93] and for block-copolymeric star molecules[94], where the rays are composed of blocks with different refractive index increment and different bond length. The various particle-scattering factors and other conformational properties are listed in tables 2 to 4.

Analytic equations were derived also for the first cumulant by using either Eq. (B.45) directly, passing to integrals and solving these integrals[82], or by using the integral representation of Eq. (B.57) and now solving these integrals[84, 93]. The final equations are, however, rather lengthy and will not be reproduced here. The integral representation of Γ is very convenient since only the particle-scattering factor is needed. The numerical result can be obtained by application of *Simpsons's rule* of numerical integration. Some of the properties of the first cumulant can, however, be calculated analytically without too much effort, and these properties will be discussed in some detail in Chap. D II.

This section may be closed with a few remarks on the rooted tree representation. It has been common practice in polymer physics to place the molecules on a special lattice in space or in plane[41-44]. An example is shown in Fig. 6 a. In such graphs, the units of the same path length are placed on shells of nearest neighbours, next nearest neighbours etc. This sort of representation has two disadvantages:

1) First, when the macromolecule is very large, it will become sometimes difficult to decide whether a unit is in the n-th or in the n+1–th shell or somewhere else. Such

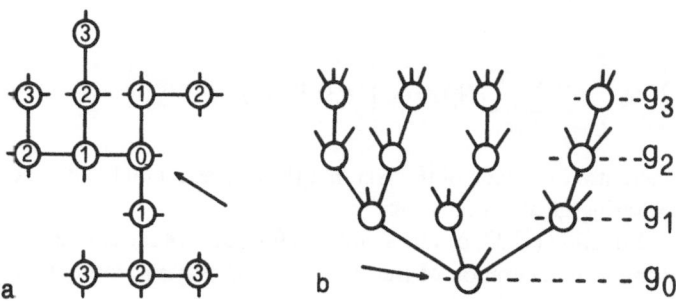

Fig. 6a, b. A tetrafunctionally branched molecule (a) placed on a lattice and (b) the corresponding rooted tree representation. Note: The units in the first, second, third etc. shell of neighbours come to lie well defined in generation g_1, g_2, g_3 etc

difficulties never exist in the rooted tree representation, because all the units of a certain shell are clearly placed here on a distinct generation.

2) Second, on placing a molecule on a special lattice, a picture is unconsciously engraved in the mind suggesting that the molecule may behave in three-dimensional space as seen in the graph or given by the computer. A special lattice always implies certain constraints which actually need not exist in this form. The rooted tree representation is free from this problem of how a molecule is embedded in space; it only displays the connectivity, and this in a very clear form[95–97].

III. Polydisperse Homopolymers

1. Some General Relationships[25–28]

Under common chemical conditions, we mostly obtain a distribution of different degrees of polymerization and of various isomeric structures. In such cases, the whole set of molecules has to be considered. In principle, for each special x-mer the structure factor, i.e. the double sum over all pairs of scattering elements has to be performed, or each monomer unit from every molecule has to be selected as the root of a tree. Thus, a full forest of trees is obtained. This forest can be ordered into groups where all the trees have the same degree of polymerization x, and for each of these groups we can in principle evaluate the double sum in the same manner as described in the previous section. The final result by summing the individual results after having weighted them according to their relative frequency. Let n_x be the number or mole fraction of x-mer in the ensemble of molecules. For each such x-mer we have x different rooted trees in the forest; thus, the relative frequency of trees of size x, i.e. with degree of polymerization x, is then x n_x/ $\sum x n_x = w_x$ which is the weight fraction of a polymer of the DP = x. Therefore, the final result of the required summation over all species of molecules may be written, using Eq. (C.14), as[34]

$$\sum_{x=1}^{\infty} n_x \left(\sum_{j}^{x} \sum_{h=0}^{\infty} \phi_n N_j(n) \right) \bigg/ \left(\sum_{x}^{\infty} n_x x \right) = \sum_{x}^{\infty} \left(n_x x \bigg/ \left(\sum_{x}^{\infty} n_x x \right) \right) \left[x^{-1} \sum_{j}^{x} \sum_{n=0}^{\infty} N_j(n) \phi_n \right]$$

or

$$\sum_{x=1}^{\infty} w_x \left[x^{-1} \sum_{j}^{x} \sum_{n=0}^{\infty} N_j(n) \phi_n \right] = P_w P_z(q^2) \tag{C.23}$$

where use was made of the fact that the expression in the brackets is x times the particle-scattering factor of an x-mer.

Equation (C.23) can be simplified further. We recall that $N_j(n)$ is the number of units in the n-th generation of tree j of a special x-mer; consequently

$$x^{-1} \sum_{j=1}^{x} N_j(n) = \langle N_x(n) \rangle \tag{C.24}$$

is the average population of units in the n-th generation of all trees of an x-mer, and furthermore

$$\sum_{x=1}^{\infty} w_x \langle N_x(n) \rangle = \langle N(n) \rangle \tag{C.25}$$

is the overall weight average population of units in the n-th generation, where the average is now over all trees in the forest. We thus find finally

$$P_w P_z(q^2) = \sum_{n=0}^{\infty} \langle N(n) \rangle \, \phi_n \tag{C.26}$$

In other words, the problem of calculating the static scattering behaviour $R(q) = Kc \, M_w P_z(q^2)$ is reduced to the determination of the average number of units in the n-th generation. This average number, which is also the average number of paths of length n issuing from a root, has then to be multiplied by the weight factor $\phi_n = \langle \exp(iq \cdot r_n) \rangle$ and the result has to be summed over all generations. Figure 7 illustrates the meaning of n_x, w_x and of the various population numbers $N_j(n)$, $\langle N_x(n) \rangle$ and $\langle N(n) \rangle$ for an ensemble of five molecular species.

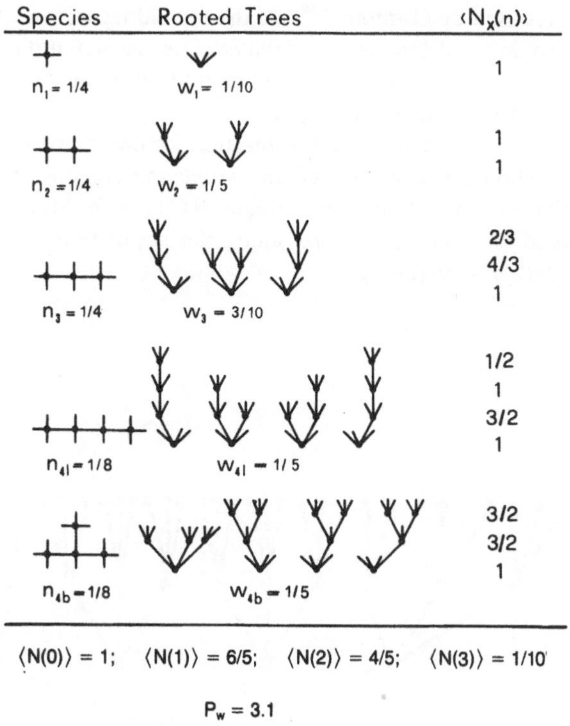

Fig. 7. Average population $\langle N_x(n) \rangle$ of units in the n-th generation for a special x-mer ($x = 1, 2, 3, 4$), and the overall average population $\langle N(n) \rangle$ of units in the n-th generation for a system with five different species. w_j denotes the weight fraction of these species in the system, and P_w is given by

$$P_w = \sum_{n=1}^{4} \langle N(n) \rangle.$$

(The number fractions n_j and some representative rooted trees have been selected arbitrarily to illustrate the calculations)

Species	Rooted Trees	$\langle N_x(n) \rangle$
$n_1 = 1/4$	$w_1 = 1/10$	1
$n_2 = 1/4$	$w_2 = 1/5$	1 1
$n_3 = 1/4$	$w_3 = 3/10$	2/3 4/3 1
$n_{4l} = 1/8$	$w_{4l} = 1/5$	1/2 1 3/2 1
$n_{4b} = 1/8$	$w_{4b} = 1/5$	3/2 3/2 1

$\langle N(0) \rangle = 1;$ $\langle N(1) \rangle = 6/5;$ $\langle N(2) \rangle = 4/5;$ $\langle N(3) \rangle = 1/10$

$P_w = 3.1$

2. Randomly Branched f-Functional Polycondensates

In this case, the polymer is formed by random polycondensation of monomers with f functional groups of equal reactivity. For each monomer selected as a root, the tree always has in its zero-th generation f functional groups which may react with another monomer to form the members of the 1-st generation. However, the units in the first generation have now only f–1 functional groups to react with further monomer units and to form the population of the 2-nd generation; the same conditions hold for the units in all higher generations. We can now construct a tree-like lattice as shown for the example of a tetrafunctional monomer in Fig. 8. Each molecular species in the ensemble can then be placed on this lattice. We now ask what is the average population (number of units) $\langle N(n) \rangle$ in the n-th generation.

Clearly, the calculation of this population and its distribution is a probabilistic problem. It is a characteristic feature of the Flory-Stockmayer theory that this probability is fully determined by the link probability or extent of reaction of the functional group. Whenever a functional group reacts, a bond is formed and the fraction of groups which have reacted, is thus called the link probability or extent of reaction, α[98]. We have now two possibilities to calculate the average number of units in the n-th generation. The first is the derivation of the *population distribution* in the various generations, from which by common techniques the average value can be calculated. This was the route followed first by Stockmayer[7] which became greatly simplified by the application of the cascade theory first used by Gordon[25–34]. In the second technique, the derivation of the distribution is avoided and the *average population* is calculated directly on the basis of *Markov chain statistics*[99–101]. We start with the application of the second, less general, but sometimes easier method of calculation.

When α is the extent of reaction for one functional group, then one monomer with f functional groups is linked on average to $f\alpha$ monomer units. Thus, the number of units in the 1-st generation is on average $\langle N(1) \rangle = f\alpha$. Since each unit in the 1-st generation can bind on average $(f-1)\alpha$ units, the population in the 2-nd generation is on average $\langle N(2) \rangle = f\alpha(f-1)\alpha$.

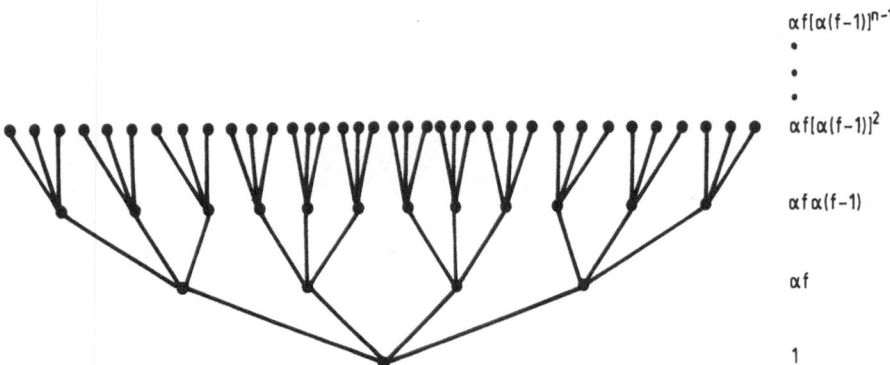

$$\alpha f [\alpha (f-1)]^{n-1}$$
$$\vdots$$
$$\alpha f [\alpha (f-1)]^{2}$$
$$\alpha f \alpha (f-1)$$
$$\alpha f$$
$$1$$

Fig. 8. The rooted tree lattice for a tetrafunctionally branched polymer, and the average population of units in the n-th generation when α was the extent of reaction of the functional groups

A unit in the 2-nd generation can again bind $(f-1)\alpha$ units, and thus

$$\langle N(3)\rangle = \langle N(2)\rangle [(f-1)\alpha] = f\alpha [(f-1)\alpha]^2, \quad \text{and finally}$$

$$\langle N(n)\rangle = f\alpha[(f-1)\alpha]^{n-1} \tag{C.27}$$

Inserting this result into Eq. (C.26) and using $\phi_n = \phi^n$, which holds for Gaussian subchains, we finally arrive at[90]

$$P_w P(q^2) = \frac{1+\alpha\phi}{1-(f-1)\alpha\phi} \cong P_w (1+u^2/3)^{-1} \tag{C.28}$$

with

$$\phi = b^2 q^2/6 ; \quad u^2 = \langle S^2\rangle_z q^2$$

$$P_w = \frac{1+\alpha}{1-(f-1)\alpha} \tag{C.29}$$

$$\langle S^2\rangle_z = b^2 f\alpha/[2(1+\alpha)(1-(f-1)\alpha)] = b^2 P_w (f-1)/2f \tag{C.30}$$

$$D_z = A \cdot [6\pi f/(f-1) bP_w]^{1/2} \tag{C.31}$$

$$\Gamma \cong q^2 D_z (1+u^2/3)^{1/2} \left[1 + \frac{u^2}{30} \frac{(1+u^2/3)^{1/2}}{(1+0.72\,u^2/3)} 3/2\right] \tag{C.32}$$

Equation (C.29) is Stockmayer's[7] famous relationship, Eq. (C.28) and (C.30) were first derived by Kajiwara et al.[34], and Eq. (C.31) and (C.32) by Burchard et al.[102].

3. Polycondensates from Monomers with Unlike Functional Groups[104, 105]

The results given for the f-functional random polycondensate, are rigorous and were obtained without any assumption or approximations. Despite this, the FS theory is now frequently quoted by physicists to be a classical example for a *mean-field theory*, which would imply a specific *mean-field approximation*[43, 103], which was, however, not invoked either by Stockmayer[7] or in the derivation of the last section. A more careful consideration of the problem reveals in fact a misunderstanding of the FS theory by physicists. Actually, a mean field approximation becomes meaningful only for monomers with unlike functional groups, or if the reactivity of unreacted groups is changed when one or more of the f-groups per monomer have already reacted. Consider, for instance, a trifunctional monomer with functional groups A, B and C. We may then introduce three different extents of reaction or link probabilities α, β and γ. Depending on the chemistry, these link probabilities will be correlated. The mean-field approximation now means that on the average these three probabilities can be replaced by an average link probability

$$\alpha \cong \beta \cong \gamma \cong \bar{a} \tag{C.33}$$

mean field approximation.

However, such a rigorous approximation need not be made in the FS theory as will now be shown with several examples. In fact, the mean-field approximation would lead in most cases to grossly incorrect results.

Here we treat in some detail the tri-functional polycondensate with three unlike functional groups[105]. In general, there will be 9 different probabilities of reaction i.e. α_1, α_2, α_3, β_1, β_2, β_3, γ_1, γ_2, γ_3 where α_1, α_2, and α_3 denote the fraction of A groups which have reacted with another A group, another B group, or another C group, respectively etc. These nine probabilities may conveniently be written as a matrix

$$\mathbf{A} \equiv \begin{bmatrix} \alpha_1 & \alpha_2 & \alpha_3 \\ \beta_1 & \beta_2 & \beta_3 \\ \gamma_1 & \gamma_2 & \gamma_3 \end{bmatrix} = \begin{bmatrix} \alpha_1 & \alpha_2 & \alpha_3 \\ \alpha_2 & \beta_2 & \beta_3 \\ \alpha_3 & \beta_3 & \gamma_3 \end{bmatrix} \tag{C.34}$$

where a symmetry relationship has been applied resulting from the fact that for a homopolymer, the number of links formed by an A-group with a B-group, must equal the number of links formed by the B-groups with an A-group. Secondly, the probabilities must fulfil the three necessary conditions

$$\begin{aligned} \alpha &= \alpha_1 + \alpha_2 + \alpha_3 \\ \beta &= \beta_1 + \beta_2 + \beta_3 \\ \gamma &= \gamma_1 + \gamma_2 + \gamma_3 \end{aligned} \tag{C.35}$$

The number of units in the n-th generation can now in principle be found in a similar manner as outlined for the random polycondensates with equal reactive groups. We first notice that the result depends on how many of the units in the n-th generation are linked with their A, or B or C groups to a unit in the preceding generation. Therefore, it will be useful to introduce a population vector $\mathbf{N}(n)$

$$\mathbf{N}(n) = (N_A(n), N_B(n), N_C(n)) \tag{C.36}$$

where the components with the indices A, B, C denote the number of units which are linked by an A, B or C group to the preceding generation $n - 1$. The average number of any type of units in the n-th generation is then the scalar product of $\mathbf{N}(n)$ with the unit vector $\mathbf{1} = (1, 1, 1)$

$$N(n) = (\mathbf{N}(n) \cdot \mathbf{1}) \tag{C.37}$$

To find the number of units in the $(n + 1)$-th generation, we have to multiply for instance $N_A(n)$ with the appropriate link probability to describe the transition from the n-th to the $(n - 1)$-th generation. For a unit which was linked by its A group to the preceding generation, there will be in general three transition probabilities p_{11}, p_{12}, p_{13}, since the two remaining groups B and C could react with an A group, or a B, or a C group of a unit in the next higher generation. Hence

$$\langle N(n+1) \rangle = \langle N(n) \rangle \mathbf{P} \tag{C.38}$$

where

$$\mathbf{P} = \begin{bmatrix} P_{11} & P_{12} & P_{13} \\ P_{21} & P_{22} & P_{23} \\ P_{31} & P_{32} & P_{33} \end{bmatrix} \tag{C.39}$$

is the *transition probability matrix*. Equation (C.38) is a recursive equation and can easily be solved to yield

$$\langle N(n) \rangle = \langle N(1) \rangle \mathbf{P}^{n-1} \tag{C.40}$$

Each component of the vector $\langle N(n) \rangle$ has now to be multiplied with the path-length dependent weighting factor. Again, we have in general three components

$$\phi_n = (\phi_{nA}, \phi_{nB}, \phi_{nC}) \tag{C.41}$$

as the path – leading from an A group in the n-th generation to the root – may be different in composition from that of a B group or a C group when the bond length between the various functional groups is different. Here we assume that on average all paths have the same composition so that for Gaussian statistics of the subchains all ϕ_n components are the same

$$\phi_{nj} = \phi_n = \exp(-b^2 q^2/6)^n$$

where b is now the avgerage-effective bond length of the various links.

Inserting Eqs. (C.40) and (C.41) into Eq. (C.26), one finally arrives at[106, 107]

$$P_w P_z(q^2) = 1 + [(\langle N(1) \rangle \cdot \mathbf{P}^{n-1} \boldsymbol{\phi}^n) \cdot \mathbf{1})]$$

or

$$P_w P_z(q^2) = 1 + (\langle N(1) \rangle \cdot [(\mathbf{1} - \mathbf{P}\,\boldsymbol{\phi})^{-1}] \cdot \mathbf{1}) \tag{C.42}$$

where we have introduced a diagonal matrix $\boldsymbol{\phi} = \mathbf{1}\,\phi$. Since $\boldsymbol{\phi}$ is diagonal, the matrix $\mathbf{P}\,\boldsymbol{\phi}$ has the same form as P but now each element is multiplied by ϕ. This form of the matrix suggests that in the general case where the bond lengths are not identical, the transition probabilities have to be multiplied by factors $\phi_{ij} = \exp(-b_{ij}^2 q^2/6)$, where b_{ij} is the bond length between functional groups i and j (i, j = A, B, C).

Comparing Eq. (C.42) with Eq. (C.28), we recognize a remarkable formal equivalence[106]. The only difference consists of the fact that the scalar transition probability $\alpha(f-1)$ for the random polycondensates of equal reactivity has to be replaced by the transition matrix $\mathbf{P}$, and the population number in the 1-st generation has to be replaced by a vector $\langle N(1) \rangle$ which contains the population of the root linked units in the first generation. Furthermore, the general Eq. (C.42) reduces exactly to the case of the random trifunctional polycondensation when all link probabilities are the same[107]. Or in other words, if a mean-field approximation of Eq. (C.33) is made, the matrix equation (C.42) reduces to the scalar form of Eq. (C.28). Eq. (C.42) demonstrates clearly that in general the branching process is a *Markovian chain process,* as it should be when the nearest neighbour effect (the correlation between the n-th and (n + 1)-th generation) is taken into account.

Now we finally have to determine the components of the population in the first generation and the various transition probabilities in the transition-probability matrix **P**. Both can be done by inspection of the graphs in Fig. 9. We thus find

$$\langle \mathbf{N}(1) \rangle = (\alpha, \beta, \gamma) \tag{C.43}$$

and,

$$\mathbf{P} = \begin{bmatrix} \beta_1 + \gamma_1 & \beta_2 + \gamma_2 & \beta_3 + \gamma_3 \\ \alpha_1 + \gamma_1 & \alpha_2 + \gamma_2 & \alpha_3 + \gamma_3 \\ \alpha_1 + \beta_1 & \alpha_2 + \beta_2 & \alpha_3 + \beta_3 \end{bmatrix} \tag{C.44}$$

The weight average degree of polymerization is found from Eq. (C.42) by setting q = 0, and the mean-square radius of gyration after differentiation of $P_z(q^2)$ with respect to q^2

$$P_w = 1 + (\langle \mathbf{N}(1) \rangle \cdot (\mathbf{1} - \mathbf{P})^{-1} \cdot \mathbf{1}) \tag{C.45}$$

$$\langle S^2 \rangle_z = -3 \; dP_z(q^2)/dq^2 \big|_{q^2 = 0} = \frac{b^2}{2 P_w} (\langle \mathbf{N}(1) \rangle \cdot (\mathbf{1} - \mathbf{P})^{-2} \cdot \mathbf{1}) \tag{C.46}$$

Finally, D_z and the first cumulant are found by integration

$$D_z = (A/\pi^{1/2}) \int_0^\infty P(q^2) \, dq \tag{C.47}$$

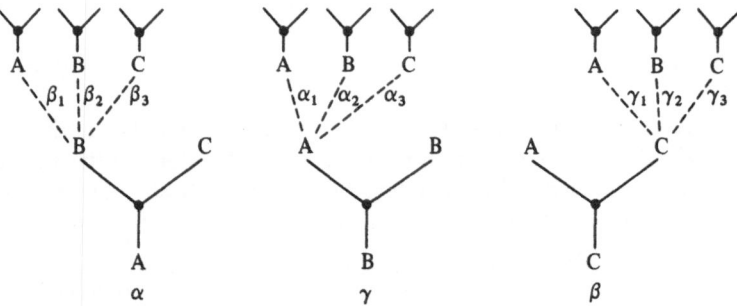

Fig. 9. The various reaction possibilities for the functional group A, B and C of a monomeric unit in the n-th generation, when group A, or B, or C were linked to a unit in the previous generation. The *Greek letters* denote the various link probabilities

and

$$\Gamma/q^2 = D_z/P_z(q^2)\left\{\int_0^\infty P(q^2+\beta^2)\,d\beta - 0.1q^2\int_0^\infty \beta^{-2}[P_z(0.72q^2)-P_z(0.72q^2+\beta^2)]\,d\beta\right\}$$

(C.48)

In principle, an analytic matrix inversion needed in Eqs. (C.42), (C.45) and (C.46) is feasible but tedious, but in general it does not give very lucid equations, while the numerical matrix inversion is readily carried out by computers. In two cases, however, the analytic inversion is worth carrying out since the resultant equations show clearly the contrast with a mean-field approximation.

The first case occurs when for chemical reasons group A can only react with group B or group C[104-108], and the second is observed when C and B are chemically identical[109-111]. The first case may be realized in glycogen of mammalia where glucose forms the monomeric units (see Fig. 10). By using the high specificity of certain enzymes, the OH-group in C1 position can be linked in a specific manner (α-glycosidic bond) to an OH-group of another monomer in the C4 position (chain formation). At the same time about 25% of all reactions produce a link to the OH-group in the C6 position (branching). In this case there are stringent restrictions on the reactions:

$$\alpha_1 = \beta_2 = \gamma_3 = \beta_3 = \gamma_2 = 0$$

(C.49)

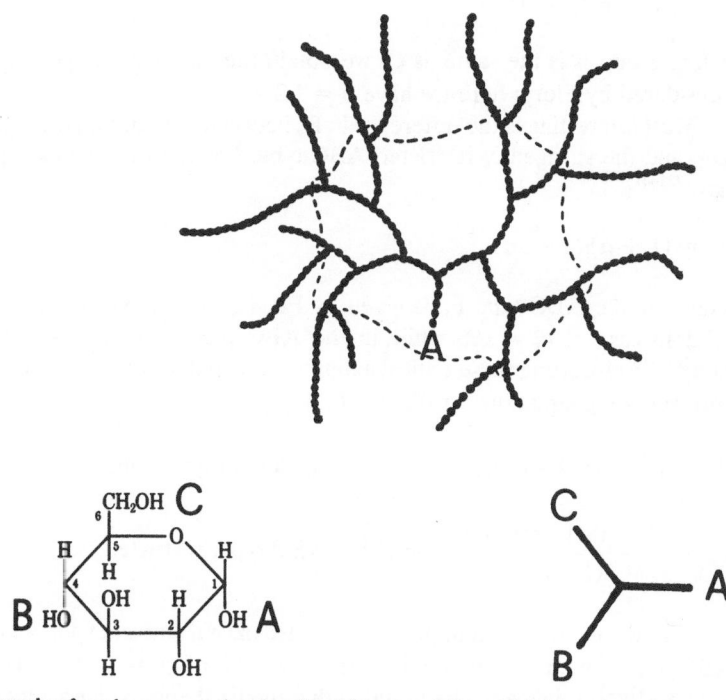

Fig. 10. Schematic graph of a glycogen molecule after Meyer and Bernfeld (*above*). The monomeric unit is the glucose (*below, left*) which may for simplicity contracted to a graph (*below, right*). Group A can react only with group B or C, all other reactions are excluded[104, 112, 113]

and

$$a_2 + a_3 = a = \beta_1 + \gamma_1 \tag{C.49}$$

or, suppressing the indices

$$\beta = a(1 - p)$$

$$\gamma = ap$$

where p is the branching probability. Performing the matrix inversion, one obtains[90, 102, 105]

$$P_w = \frac{1 - a^2[(1 - p)^2 + p^2]}{(1 - a)^2} \tag{C.51}$$

$$P_z(q^2) = P_w^{-1} \frac{1 - a^2\phi^2[(1 - p)^2 + p^2]}{(1 - a\phi)^2} \tag{C.52}$$

$$\langle S^2 \rangle_z \equiv b^2 P_w(1 + 2B)/(4(1 + B)^2) \tag{C.53}$$

$$D_z = (A/(6\pi)^{1/2})(2 + B)/(2(1 + B)P_w)^{1/2} \tag{C.54}$$

where

$$B = p(1 - p)/(1 - a) \tag{C.55}$$

When group B is the same as C, we obtain the AB_2-type polymerisate which was first considered by Flory; here we have p = 1/2.

Most interesting is the difference in P_w between the random trifunctional polycondensate and the stringently restricted ABC type. For the tri-functional polycondensate we have[7, 104]

$$P_w = (1 + a)/(1 - 2a)$$

while for the ABC type P_w is given by Eq. (C.51)[104]. We see that P_w diverges in the random case at $a_c = 1/2$, while in the ABC type only at $a_c = 1$ i.e. at full conversion[104, 110]. Inserting these critical extents of reactions, where gelation takes place, in the corresponding equations for P_w, we find

$$P_w = (1 + a)/(1 - a/a_c) \quad \approx \varepsilon^{-1} \quad \text{random polycondensate}$$

$$P_w = \frac{1 - a^2|(1 - p)^2 + p^2|}{(1 - a/a_c)^2} \approx \varepsilon^{-2} \quad \text{ABC type, restricted}$$

In other words, the random polycondensates have a critical exponent of $\gamma_c = 1$ while the ABC type under restriction has an exponent of $\gamma_c = 2$. Percolation theory yields $\gamma_c = 1.8$[43, 44]. It has often been stated that the FS theory is characterized by a mean-field critical exponent of $\gamma_c = 1$, but now we see that the FS theory is more flexible and shows critical exponents between 1 and 2.

In general, the critical point of gelation is defined by $P_w \to \infty$[7], which yields with Eq. (C.45) the very compact relationship[105-107]

$$|1 - P| = 0 \qquad\qquad\qquad (C.56)$$

IV. The Cascade Procedure[25-28, 34, 106, 114, 115]

1. Some General Remarks

The direct method for evaluation of $\langle N(n) \rangle$ could be applied in the last two examples of polycondensation with not too great a difficulty. For more complex systems, however, as for instance the formation of branched epoxide resins[115], or for branching processes which involve chain reactions[106, 107, 116, 117], this technique becomes very difficult and non-transparent so that errors can scarcely be avoided, mostly by relevant terms being left out. The evaluation of the answer is on the other hand greatly facilitated by the use of probability-generating functions[118] since the desired averages of the distribution are obtained by a simple differentiation operation on the probability-generating function. Furthermore, the rules for calculation with generating functions are simple and very efficient so that the final generating function of a complex statistical problem can be obtained by only a few operations with the generating functions of elementary processes. Finally, the probability-generating function contains the same information as the corresponding probability distribution itself. In fact, the distribution can be derived from the corresponding-generating function although in many cases only approximate formulae can be obtained.

Since probability-generating functions are not often used in polymer science, we first give a brief outline of the properties before applying them to the problems of branched polymers.

2. Probability Generating Functions (pgf)[118]

a) Definition

Let w_x be a probability distribution (e.g. the weight fraction of a polymer) of finding an event (the degree of polymerization), where x is an integer. Then the corresponding generating function $W(s)$ is defined as the sum over all the events w_x multiplied by a variable s raised to the x-th power

$$W(s) \equiv \sum_{x=1}^{\infty} w_x s^x \qquad\qquad\qquad (C.57)$$

where

$$0 \leq s \leq 1$$

and s may be real or complex (s is usually only a dummy or auxiliary variable). In some cases, however, the variable s is replaced by a function $f(s)$, whereby Eq. (C.57) keeps its meaning of a generating function.

b) Some Properties of the Probability Generating Function

The pgf is evidently a transformation of the distribution from a discrete space x into a continuous space s, and this has the advantage that the pgf can be treated in the same way as other analytic functions (which is not the case for the distribution). Making use of this advantage, we find the following main properties:

(1) Normalization conditions

$$W(s)_{s=1} \equiv W(1) = \sum_{x=1}^{\infty} w_x = 1 \tag{C.58}$$

(2) Mean value of the distribution

$$\partial W/\partial s|_{s=1} \equiv W'(1) = \sum_{x=1}^{\infty} x w_x = \langle x \rangle \tag{C.59}$$

For instance, if w_x is the weight fraction of polymerization degree x, the quantity $\langle x \rangle = \langle x \rangle_w = P_w$ is the weight average degree of polymerization.

c) Moments of Distribution

More generally, one finds for the k-th moment M_k of the probability distribution

$$M_k = \partial^n W(s)/\partial \ln s^n|_{s=1} \tag{C.60}$$

In particular, when w_x is the weight fraction of polymers, one has

$$P_z = (M_2/M_1) + 1 \tag{C.61}$$

where P_z is the z-average degree of polymerization.

Number and weight fraction generating functions[31, 119]
Let

$$H(s) \equiv \sum_{x=1}^{\infty} h_x s^x$$

$$W(s) \equiv \sum_{x=1}^{\infty} w_x s^x$$

be the frequency (number fraction) and the weight fraction generating functions, respectively, where

$$w_x = xh_x / \sum_{x=1}^{\infty} xh_x = xh_x / P_n \qquad (C.62)$$

Then $W(s)$ can be derived from $H(s)$ by a differentiation operation and $H(s)$ from $W(s)$ by an integration operation

$$W(s) = sH'(s)/H'(1) \qquad (C.63)$$

$$H(s) = H'(1) \int_0^s W(t)/t \, dt \qquad (C.64)$$

Thus, from Eq. (C.64)

$$1/P_n = \int_0^1 [W(t)/t] \, dt \qquad (C.65)$$

where $P_n = H'(1)$ is the number average degree of polymerization. Note that $H(1) = 1$, because of the normalization condition of a probability distribution.

Example:
These properties may be illustrated using the example of the familiar, most probable distribution[120-125] for linear chains. The frequency distribution for polymers with a degree of polymerization x can be written in terms of the link probability α

$$h_x = (1 - \alpha) \alpha^{x-1} \qquad (C.66)$$

Hence[125]

$$H(s) = (1 - \alpha)s \sum_{x=1}^{\infty} (\alpha s)^{x-1} = (1 - \alpha)s/(1 - \alpha s) \qquad (C.67)$$

$$H'(1) = P_n = (1 - \alpha)^{-1} \qquad (C.68)$$

$$W(s) = (1 - \alpha)^2 s/(1 - \alpha s)^2 \qquad (C.69)$$

$$W'(1) = (1 + \alpha)/(1 - \alpha) = P_w \qquad (C.70)$$

Equation (C.68) and Eq. (C.70) are well known averages for the most probable distribution.

d) Calculation Rules for pgf-s

(1) The Convolution Theorem. Let $w_1(x)$ and $w_2(y)$ be two probability distributions. Often one is interested in the distribution of finding the sum $z = x + y$. Such a case occurs in polymer science, for instance, in the radical polymerization with termination by radical combination[126-128]. The resulting probability distribution is given by

$$w(z) = \sum_{\zeta=0}^{z} w_1(z - \zeta) w_2(\zeta) \equiv w_1 * w_2 \tag{C.71}$$

This type of coupling occurs quite often in probability problems and is called a convolution of two distributions (mostly abbreviated by a star product). Passing to generating functions, the rather complicated sum of products of probabilities is transformed into a simple product of the corresponding pgf-s.

$$W(s) = W_1(s) W_2(s) \tag{C.72}$$

Equation (C.72) is known as the *convolution theorem*.

(2) The Compound Distribution and Cascade Substitution. Another type of composed distribution may be constructed when the number of participants is not fixed but follows a distribution of the sums. The resultant distribution is called a *compound distribution*. For illustration, the distribution of units in the n-th generation for a f-functional polycondensate may be considered (see Fig. 11).

Let $N_i(1)$ be the distribution of units in the first generation $(i = 1, 2, \ldots f)$, i.e. the probability to find i-units in the first generation. We wish to know the distribution $N_j(2)$ of units in the second generation. When we start from a unit in the first generation, then $N_{kj}(2)$ may be the distribution of "offspring" for the k-th of them. In a case, where – of the f-functional groups of the root – only exactly i-groups have reacted, the population distribution in the second generation would be the i-fold convolution of the $N_{kj}(2)$ distribution if we assume the same distribution for all k, f-functional groups, i.e.

$$N_j^{(i)}(2) = [N_{kj}(2)]^{i*} \equiv N_{1j}(2) * N_{2j}(2) * \ldots * N_{ij}(3) \tag{C.73}$$

Actually, however, the probability of finding exactly i-units in the first generation is $N_i(1)$, and the total distribution is therefore

$$N_j(2) = \sum_{i=0}^{f} N_1(1) N_j^{(i)}(2) = \sum_{i=0}^{f} N_i(1) [N_{kj}(2)]^{i*} \tag{C.74}$$

Note: $N_i(1)$ is the probability that the root has i-"offspring", and $N_{kj}(2)$ is the probability for j-"offspring" of a unit in the first generation.

We introduce now generating functions for the "offspring"[25]

$$F_0(s) = \sum_{i=0}^{f} N_i(1) s^i \tag{C.75 a}$$

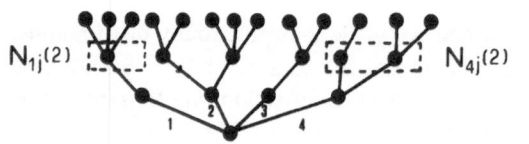

$N_{1j}(2)$ $N_{4j}(2)$

3
2
1
0

Fig. 11. A special rooted tree as illustration for the connection between the total distribution of units in the second generation $N_j(2)$ and those arising from the various functional groups 1, 2, 3 and 4 denoted as $N_{1j}(2)$, $N_{2j}(2)$ etc

$$F_1(s) = \sum_{j=0}^{f-1} N_{kj}(2) s^j \qquad (C.75\,b)$$

and for the population distribution in the first two generations

$$G_1(s) = \sum_{j=0}^{f} N_j(1) s^j = F_0(s)$$

$$G_2(s) = \sum_{i=0}^{\infty} N_i(2) s^i$$

Now multiplying Eq. (C.74) on both sides by s^i and summing over all i, we obtain after application of the convolution theorem

$$G_2(s) = \sum_{i=0}^{f} N_i(1) [F_1(s)]^i$$

or

$$G_2(s) = F_0(F_1(s)) \qquad (C.76)$$

which is easily verified when $F_1(s)$ is taken as an auxiliary function, say s^*.

Proceeding stepwise to higher generations, one finally obtains for the pgf of the n-th generation

$$G_n(s) = F_0(F_1(\ldots F_{n-1}(s))) \qquad (C.77)$$

Thus, whenever the pgf for the "offspring" of units in the various generations is known, the final distribution for the population in the n-th generation is then obtained by a cascade substitution of the kind given by Eq. (C.77). The cascade substitution is often called a *Galton-Watson process* since this theorem was first derived by Watson and Galton[16–18]. The average number in the various generations is obtained by differentiation with respect to s at $s = 1$, which yields

$$\left. \frac{\partial G_n(s)}{\partial s} \right|_{s=1} = \langle N(n) \rangle = F_0'(1) [F_1'(1)]^{n-1} \qquad (C.78)$$

where we have made use of the fact that the "offspring" distribution is the same for all generations larger than the zero-th, i.e. $n = 1, 2$, etc. Note that the root has f functional groups while in all other generations the units have only $f - 1$ functional groups, which could lead to the next higher generation.

Finally, when we sum over all generations, we obtain the weight average degree of polymerization

$$P_w = F_0'(1) \sum_{n=1}^{\infty} F_1'(1)^{n-1} + 1 = [F_0'/(1 - F_1'(1))] + 1 \qquad (C.79)$$

This relationship was first derived by Gordon[25].

The angular dependence of the static scattering is given quite generally by Eq. (C.26)

$$P_w P_z(q^2) = \sum_{n=1}^{\infty} \langle N(n) \rangle \phi_n + 1 \tag{C.26}$$

Hence, with Eq. (C.78) we obtain[34]

$$P_w P_z(q^2) = 1 + F_0'(1) \sum_{n=1}^{\infty} F_1'(1)^{n-1} \phi_n$$

$$= 1 + F_0'(1) \phi/(1 - F_1'(1) \phi) \tag{C.80}$$

where the second relationship is valid only for Gaussian subchains, where $\phi_n = \phi^n$. All other quantities can be derived as outlined before.

Before continuing the theory, it is worth while to stop for a moment and to consider the derivation of the final formula. To obtain a relationship for P_w, it was necessary to derive a generating function for the population in the n-th generation. This pgf could be expressed in a simple manner in terms of the "offspring"-generating function $F_0(s)$ and $F_1(s)$ for the root and members in the first generation. Thus, in the final result only the derivatives of the two generating functions $F_0(s)$ and $F_1(s)$ appear. These generating functions can now very easily be derived by simple considerations using the definition of a generating function.

The construction of $F_0(s)$ and $F_1(s)$ may be demonstrated with the simple example of the random f-functional polycondensation[25, 31, 34, 131].

First, we consider one functional group. The distribution of reactants is very simple in this case because there is only one probability α (the functional group has bound another monomer unit) and a probability $1 - \alpha$ (no unit is bound). The definition of a generating function requires that the probability for no event has to be multiplied by $s^0 = 1$, and the probability for one event by $s^1 = s$; hence, the p.g.f for one functional group A

$$F_A(s) = 1 - \alpha + \alpha s \tag{C.81}$$

The monomer unit chose as a root, however, bears f functional groups, and therefore the distribution of offspring is the f-fold convolution of the functional-group probability; or applying the convolution theorem (C.IV.c)

$$F_0(s) = (1 - \alpha + \alpha s)^f \tag{C.82}$$

A unit in the first generation, or in any higher generation, has only $(f - 1)$ functional groups available for further reaction, thus

$$F_1(s) = (1 - \alpha + \alpha s)^{f-1} = F_n(s) \tag{C.83}$$

When these generating functions are introcuced into Eq. (80), one obtains

$$P_w P_z(q^2) = 1 + \alpha f \phi/[1 - \alpha(f-1)\phi] = (1 + \alpha\phi)|1 - \alpha(f-1)\phi|^{-1} \tag{C.84}$$

which is the result already derived in C.III.2.

3. Weight and Path-Weight Generating Functions

The derivation of Eq. (C.84) can even more be simplified by introducing a weight[25], or more general, a path-weight generating function[34]. The weight-generating function $W(s)$ is the probability-generating function for the weight fraction w_x of polymers with a DP of x, such that by differentiation the weight average degree of polymerization can be found (see Eq. C.57 and C.59). This weight-generating function can be obtained from the generating functions for the offspring in the various generations $F_0(s)$, $F_1(s)$ etc. by the following infinite cascade substitution[25, 26]:

$$W(s) = sF_0(sF_1(sF_1...)) \tag{C.85}$$

or alternatively

$$W(s) = sF_0(U) \tag{C.85a}$$

$$U(s) = sF_1(U) \tag{C.85b}$$

In fact, differentiation with respect to s at s = 1 yields

$$W'(1) = 1 + F_0'(1)U'(1)$$
$$U'(1) = 1 + F_1'(1)U'(1)$$

or

$$W'(1) = 1 + F_0'(1)/(1 - F_1'(1)) = P_w \tag{C.79}$$

which is again the result obtained before.

Good's method[26, 27, 130] can be extended by weighting each offspring-generating function by $s\phi^n$, where ϕ_n is a special weighting function of the generation. With this weighting the various generations now become different, while in the weight-generating function each generation n > 0 had the same weight 1. The resultant generating function

$$U_0(s) = s^{\phi_0}F_0(U_1)$$
$$U_1(s) = s^{\phi_1}F_1(U_2)$$
$$............$$
$$U_n(s) = s^{\phi_n}F_n(U_{n+1}) \tag{C.86}$$
$$............$$

will be called a path-weight generating function[34] for the following reason: differentiation of $U_0(s)$ at s = 1 yields an average which can be recognized when the differentiation is carried out. We first obtain

$$U_0'(1) = \phi_0 + F_0'(1)U_1'(1)$$
$$U_1'(1) = \phi_1 + F_1'(1)U_2'(1)$$
$$................$$
$$U_n'(1) = \phi_n + F_1'(1)U_{n+1}'(1) \tag{C.87}$$

After iterative substitution, one obtains

$$U_0'(1, \phi) = \phi_0 + F_0'(1) \sum_{n=1}^{\infty} [F_1'(1)]^{n-1} \phi_n \tag{C.88}$$

However, $F_0'(1)[F_1'(1)]^{n-1}$ is the average number of units in the n-th generation $\langle N(n) \rangle$ (see Eq. (C.78)), and thus

$$U_0'(1, \phi) = 1 + \sum_{n=1}^{\infty} \langle N(n) \rangle \phi_n = P_w \langle \phi \rangle_z \tag{C.89}$$

In other words, each generation is weighted after differentiation with a function that depends on the path length. One easily verifies Eq. (C.84) for the random polycondensates when $\phi_n = \phi^n$ (Gaussian statistics) is assumed.

The advantage of the infinite cascade of this kind of path-weight generating function consists of the possibility of calculating conformational properties in one step instead of the two steps used before, where first the number of units in the generations has to be calculated, and second the result has to be weighted and summed over all generations. The generating functions are given here in an implicit form, and the explicit form may indeed be rather complicated, but is not needed for the calculation of the averages.

We have already mentioned that a generating function contains the same information as the corresponding probability distribution, and the distribution can be obtained by special expansion techniques from the p.g.f[25, 27-29, 83, 84, 132-134]. The same technique of expansion can also be applied to the path-weight generating function and yields the properties of molecularly monodisperse fractions of the branched polymer. We will give an example later in Chap. D.I. (See also Appendix).

4. Unlike Functional Groups

To complete the theoretical framework, we treat now characteristic examples which also demonstrate the full power of the cascade theory. This case is once again the polycondensation of trifunctional monomers with unlike functional groups[105]. Another example is the random co-condensation of an f-functional monomer RA_f and a bifunctional RB_2 monomer, which is treated in the next chapter. All other cases can in principle be reduced to these general examples.

The homopolycondensation of monomers with unlike functional groups has been treated already in Chap. C.III.2, where we also have shown that for an exhaustive description of the number $\langle N(n) \rangle$ in the n-th generation, we have to consider the cases where either group A, or group B, or group C was linked to the n − 1 th generation. It will be natural, therefore, to introduce a vector-matrix notation into the theory of cascade processes. The formalism of constructing the path-weight generating function still remains the same. This means that we have to go through three stages:
(i)   setting up the probability-generating functions for the various generations F_0, F_1 etc.
(ii) performing the infinite cascade substitution similar to Eq. (C.86), and finally
(iii) carrying out the differentiation with respect to s at s = 1, and summing the result to give the desired average.

(i) The p.g.f for the Offspring Number in the Various Generations

Since there are three different functional groups, the corresponding probability-generating functions for the reaction of these groups will also be different. These are easily found with the reactivity matrix A given by Eq. (C.34)

$$F_A(s) = 1 - \alpha + \alpha_1 s_1 + \alpha_2 s_2 + \alpha_3 s_3$$
$$F_B(s) = 1 - \beta + \beta_1 s_1 + \beta_2 s_2 + \beta_3 s_3 \qquad (C.90)$$
$$F_C(s) = 1 - \gamma + \gamma_1 s_1 + \gamma_2 s_2 + \gamma_3 s_3$$

where

$$s = (s_1, s_2, s_3)$$

is a vector of labelled auxiliary variables in which the components denote the reaction with group A, (s_1), with group B, (s_2), or group C, (s_3) respectively.

The generating functions for the offspring of a unit in the zero-th generation $F_0(s)$, follows from the convolution of the three distributions of the individual functional groups, or in terms of generating functions

$$F_0(s) = F_A(s) F_B(s) F_C(s) \qquad (C.91)$$

For the units in the first generation, the result is different when group A is similar to the preceding generation, or group B, or group C. Applying the convolution theorem, one finds by inspection of Fig. 9

$$F_{1a}(s) = F_B(s) F_C(s)$$
$$F_{1b}(s) = F_A(s) F_C(s) \qquad (C.92)$$
$$F_{1c}(s) = F_A(s) F_B(s)$$

Thus, the p.g.f. for the offspring of units in the first generation is most conveniently written as a vector

$$F_1(s) = (F_{1a}(s), F_{1b}(s), F_{1c}(s)) \qquad (C.93)$$

and again we have

$$F_n(s) = F_1(s)$$

for all further generations.

(ii) The Cascade Substitution

In the random polycondensation process with f like functional groups, this was achieved by multiplying $F_n(s)$ by s^{ϕ_n} and substituting s in the n-th p.g.f. by a function $U_{n+1}(s)$.

Since in the present case s is a vector, the functions $U_n(s)$ must also be vectors. Hence, the path-weight generating functions are

$$U_0(s) = F_0(U_1) \tag{C.94a}$$

$$U_1(s) = F_1(U_2) \tag{C.94b}$$

. . . .

$$U_n(s) = F_1(U_{n+1})$$

Explicitly written, this yields

$$U_0(s) = s^{\phi_0}(1 - \alpha + \alpha_1, U_{1a} + \alpha_2 U_{1b} + \alpha_3 U_{1c})$$

$$\times (1 - \beta + \beta_1 U_{1a} + \beta_2 U_{1b} + \beta_3 U_{1c})(1 - \gamma + \gamma_1 U_{1a} + \gamma_2 U_{1b} + \gamma_3 U_{1c}) \tag{C.95a}$$

$$U_{1a}(s) = s^{\phi_1}(1 - \beta + \beta_1 U_{2a} + \beta_2 U_{2b} + \beta_3 U_{2c})(1 - \gamma + \gamma_1 U_{2a} + \gamma_2 U_{2b} + \gamma_3 U_{2c})$$

$$U_{1b}(s) = s^{\phi_1}(1 - \alpha + \alpha_1 U_{2a}\alpha_2 U_{2b} + \alpha_3 U_{2c})(1 - \gamma + \gamma_1 U_{2a} + \gamma_2 U_{2b} + \gamma_3 U_{2c}) \tag{C.95b}$$

$$U_{1c}(s) = s^{\phi_1}(1 - \alpha + \alpha_1 U_{2a} + \alpha_2 U_{2b} + \alpha_3 U_{2c})(1 - \beta + \beta_1 U_{2a} + \beta_2 U_{2b} + \beta_3 U_{2c})$$

(iii) The Average obtained from the Path-Weight Generating Functions

The average is obtained as the solution of

$$U_0'(1, \phi) = P_w \langle \phi \rangle_z$$

where the differentiation is defined as

$$dU_0/ds \equiv \frac{\partial U_0}{\partial U_{1a}} \cdot \frac{dU_{1a}}{ds} + \frac{\partial U_0}{\partial U_{1b}} \cdot \frac{dU_{1b}}{ds} + \frac{\partial U_0}{\partial U_{1c}} \cdot \frac{dU_{1c}}{ds} \tag{C.96}$$

and where the abbreviation d/ds means differentiation with respect to the individual components of s, i.e.

$$d/ds = \partial/\partial s_1 + \partial/\partial s_2 + \ldots$$

Performing the differentiation, one finds from Eq. (C.94a)

$$U_0'(1) = \phi_0 + \langle N(1) \rangle U_1'(1) \tag{C.97a}$$

and for the other generations

$$U_1'(1) = \phi_1 + P U_2'(1)$$

$$U_2'(1) = \phi_2 + P U_3'(1) \tag{C.97b}$$

. . . .

where $\langle N(1) \rangle$ contains as components the factors $\partial F_0(1)/\partial U_{1a}$, $\partial F_0(1)/\partial U_{1b}$ and $\partial F_0(1)/\partial U_{1c}$; and these are the coefficients of the components of $U_1'(1)$ when written in component form. Since $U_1(s)$ is already a vector and the differentiation of each of the components yields a vector, we now obtain for $U_1'(1)$ a coefficient matrix P to the vector $U_2'(1)$. This can easily be verified from Eq. (C.95). Furthermore,

$$\langle N(1) \rangle = (\alpha, \beta, \gamma) \tag{C.43}$$

$$P = \begin{bmatrix} \beta_1 + \gamma_1 & \beta_2 + \gamma_2 & \beta_3 + \gamma_3 \\ \alpha_2 + \gamma_1 & \alpha_2 + \gamma_2 & \alpha_3 + \gamma_3 \\ \alpha_1 + \beta_1 & \alpha_2 + \beta_2 & \alpha_3 + \beta_3 \end{bmatrix} \tag{C.44}$$

These are equations which were already derived in Chap. C.III.2.

The recursive equation can now be solved step by step and yields the final result

$$U_1'(1) = \sum_{n=1}^{\infty} P^{n-1} \phi_n = \sum_{n=1}^{\infty} (P\phi)^{n-1} \phi \cdot 1 \tag{C.42 a}$$

and

$$U_0'(1) = P_w \langle \phi \rangle_z = \phi_0 + [\langle N(1) \rangle \phi \cdot (1 - P\phi)^{-1} \cdot 1] \tag{C.42 b}$$

The advantage of the cascade process compared with the method applied in Chap. C.III.2 is that the probability matrix P follows immediately from the generating function formalism as being the coefficient matrix to the components of the next higher generation, and this makes the role of a transition matrix immediately clear.

V. General Copolymers

1. The Static Scattering Function[28, 60, 61]

The scattering theory of linear copolymers was first developed by Benoit and his coworkers[60, 61]. Later, Gordon and his coworkers applied the cascade mechanism to branched copolymers and succeeded in the calculation of the molecular weight averages M_w and M_n, the molecular weight distribution, and the mean-square radius of gyration $\langle S^2 \rangle_z$[28]. In this treatment, however, the possibility was neglected to treat differing refractive index increments ν_A and ν_B for the two monomers in a binary copolymer. In the following both theories are combined.

The general formula for the static scattering was given by Eq. (B.19)

$$\frac{R(q)}{Kc} = M_p P_z(q^2) = \sum_{x_i}^{\infty} \left(\frac{w_{xi}}{\nu^2 M_{xi}} \right) \sum_{j}^{x_i} \sum_{k}^{x_i} \phi_{jk} M_{oj} M_{ok} \nu_j \nu_k \tag{B.19}$$

This complicated triple sum can be simplified by applying the rooted tree treatment. Here the outline is confined to a binary copolymer; the results may easily be generalized to copolymers with r components.

In the binary copolymer the two molecular weights of the monomeric units are M_{oA} and M_{oB}, the corresponding refractive index increments ν_A and ν_B, and the overall composition (mole fractions) n_A and n_B. This overall composition will in general differ from the composition of individual molecules. Thus, the molecular weight M_{xi} of the i-th isomer with polymerization degree x is given by

$$M_{xi} = x\,(n_{Ai}M_{oA} + n_{Bi}M_{oB}) \tag{C.98}$$

where n_{Ai} and n_{Bi} are the mole fractions of the two monomers in this special isomer. Selecting now each monomeric unit in the total ensemble of molecules as the root of a tree, we obtain a forest which can be ordered into subforests with trees having either A or B units as the root. Each of these subforests can then be subdivided into groups of trees with the same degree of polymerization and identical composition. We now follow the line developed in chapter III.1 and obtain in the first stage

$$M_{app}P_z(q^2) = \nu^{-2} \sum w_{xi} \left[\left(\frac{M_{oA}n_{Ai}A}{n_{Ai}M_{oA} + n_{Bi}M_{oi}} \right) \frac{1}{n_{Ai}x} \sum_{j=1}^{xn_{ai}} \sum_{n=0}^{x} N_j(n)\,\phi_n M_{on}\nu_n \right. \tag{C.99}$$

$$\left. + \left(\frac{M_{oB}n_{Bi}B}{n_{Ai}M_{oA}\,x\,n_{Bi}M_{oB}} \right) \frac{1}{n_{Bi}x} \sum_{k=1}^{xn_{bi}} \sum_{n=0}^{x} N_k(n)\,\phi_n M_{on}\nu_n \right]$$

where the index j runs over all A-rooted trees and k over all B-rooted trees with DP = x. Clearly, the sums

$$\frac{1}{n_{Ai}x} \sum_{j=1}^{xn_{Ai}} N_j(n) = \langle N_{xi}(n)\rangle_A \tag{C.100 a}$$

$$\frac{1}{n_{Bi}x} \sum_{k=1}^{xn_{Bi}} N_k(n) = \langle N_{xi}(n)\rangle_B \tag{C.100 b}$$

are the averages of the population of units (A and B) in the n-th generation for the A-rooted and B-rooted trees of DP = x for all special isomers. Insertion of these equations into Eq. (C.99) yields

$$M_{app}P_z(q^2) = \nu^{-2} \left[m_A\nu_A \sum_{n=0}^{\infty} \langle N(n)\rangle_A \phi_n M_{on}\nu_n + m_B\nu_B \sum_{n=\nu}^{\infty} \langle N(n)\rangle_B \phi_n M_{on}\nu_n \right] \tag{C.101}$$

where

$$\sum w_{xi} \langle N_{xi}(n) \rangle_A = \langle N(n) \rangle_A \qquad\qquad\qquad (C.102\,a)$$

$$\sum w_{xi} \langle N_{xi}(n) \rangle_B = \langle N(n) \rangle_B \qquad\qquad\qquad (C.102\,b)$$

denote the average populations of monomeric units in the n-th generation of all A-rooted trees, respectively, in the ensemble (see Fig. 12). Equations (C.101) may be compared with Eq. (C.26) for a homopolymer. (Note: for v_n and M_{on} in Eq. (C.101), the refractive index increment and the monomeric unit molecular weight of that type which forms the end of a path in the n-th generation has to be used.)

The main problem is now the derivation of the average population in the n-th generation for the A-rooted and B-rooted trees. This is easily achieved by the use of generating functions (g.f.). Evidently, two sets of generating functions are needed for offspring in the various generations: $F_{oA}(s)$, $F_{nA}(s)$, and $F_{oB}(s)$, $F_{nB}(s)$, where the vector of auxiliary variables $s = (s_A, s_B)$ has two components which describe whether a unit is linked to a A-unit or to a B-unit. The derivatives of the generating functions at $s = 1$ gives the number of offspring for each of the units in the n-th generation. The probability-generating function for the path weight g.f. is obtained from a cascade substitution in exactly the same way as was outlined in Eq. (C.86). Hence

$$U_{oA}(s) = s^{\hat{\phi}o} F_{oA}(U_{1A}, U_{1B})$$

$$U_{oB}(s) = s^{\hat{\phi}o} F_{oB}(U_{1A}, U_{1B})$$

$$\qquad\qquad\qquad\qquad\qquad (C.103)$$

$$U_{nA}(s) = s^{\hat{\phi}n} F_{nA}(U_{n+1,A}, U_{n+1,B})$$

$$U_{nB}(s) = s^{\hat{\phi}n} F_{nB}(U_{n+A}, U_{n+1,B})$$

where again the zero-th generation has to be distinguished from the higher generations, and where the abbrevation

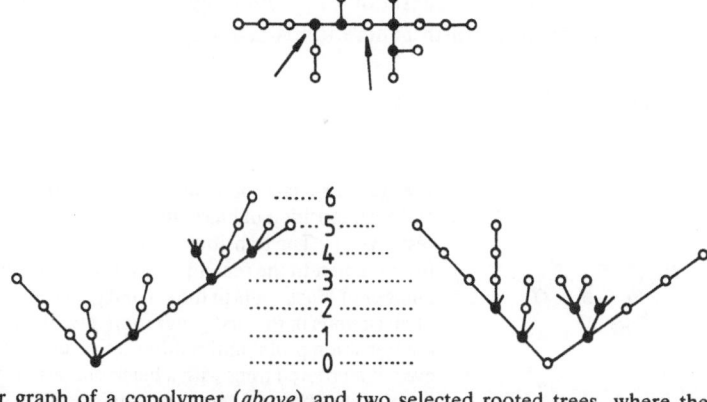

Fig. 12. The molecular graph of a copolymer (*above*) and two selected rooted trees, where the *arrows* denote the unit selected as roots

$$\hat{\phi}_n \equiv \phi_n v_n M_{on}$$

was used.

Differentiation with respect to $s = (s_A, s_B)$ at $s_A = 1$ and $s_B = 1$ yields after some rearrangements

$$U'_o(1) = \hat{\phi}_o + \langle N(1)\rangle \sum_{n=1}^{\infty} P^{n-1}\hat{\phi}_n \tag{C.104}$$

where use has been made of the vector-matrix notation with the matrix $\langle N(1)\rangle$ of the units in the first generation

$$\langle N(1)\rangle = \begin{bmatrix} \langle N(1)\rangle_{AA} & \langle N(1)\rangle_{AB} \\ \langle N(1)\rangle_{BA} & \langle N(1)\rangle_{BB} \end{bmatrix} \tag{C.105}$$

and the transition-probability matrix P

$$P = \begin{bmatrix} P_{AA} & P_{AB} \\ P_{BA} & P_{BB} \end{bmatrix} \tag{C.106}$$

where the elements of $\langle N(1)\rangle$ and P are defined as

$$\langle N(1)\rangle_{AA} = \partial F_{oA}/\partial s_A; \quad \langle N(1)\rangle_{AB} = \partial F_{oB}/\partial s_B$$
$$\langle N(1)\rangle_{BA} = \partial F_{oB}/\partial s_A M; \quad \langle N(1)\rangle_{BB} = \partial F_{oB}/\partial s_B \tag{C.107}$$

and

$$P_{AA} = \partial F_{nA}/\partial s_A; P_{AB} = \partial F_{nA}/\partial s_B \quad \text{etc.} \tag{C.108}$$

$\langle N(1)\rangle_{AA}$ and $\langle N(1)\rangle_{AB}$ are the units of type A and B, respectively, which are linked to an A-root; similar expressions exist for the two other components. Likewise the transition probabilities p_{AA}, p_{AB} etc. denote the number of A-units or B-units in the n-th generation, which are linked to an A-unit in the n-th generation etc. (Fig. 13).

In a similar manner – outlined in the derivation of Eq. (C.78) for the average population of units in the n-th generation of a homopolymer – it can be shown that the number of A-monomers in the n-th generation is given by

$$\langle N(n)\rangle_A = \langle N(1)\rangle_A \cdot P^{n-1} \cdot 1 \tag{C.109}$$

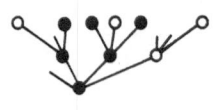

$P_{bb} = 3/2$; $P_{bw} = 1$;

$P_{wb} = 0$; $P_{ww} = 1/2$;

Fig. 13. A special copolymer tree, rooted on a black unit, of DP = 10, and a composition of black and white units of $n_b = 0.6$ and $n_w = 0.4$, respectively. The transition probabilities refer here to the transition from the first to the second generation. For example, p_{bb} is the number of black units in the second generation divided by the number of black units in the first generation, etc. In individual rooted trees, the transition probabilities differ from generation to generation. However, if all rooted trees with a black unit are considered, the transition probabilities become on average independent of the generation, if $n > 0$. The same holds for all white unit rooted trees

and

$$\langle N(1) \rangle_A = (\langle N(1) \rangle_{AA}, \langle N(1) \rangle_{AB}) \tag{C.110}$$

with a similar expression for $\langle N(1) \rangle_B$, and $\mathbf{1}$ is the unit column vector. Comparing now Eq. (C.104) with Eq. (C.101) and making use of Eq. (C.109), we notice that the scattering equation can be written as

$$M_{app}P_z(q^2) = U_o'(\mathbf{1}) = \nu^{-2}(\hat{\mathbf{m}} \cdot U_o'(\mathbf{1})) \tag{C.111}$$

where the vector $\hat{\mathbf{m}}$ contains the two components $\hat{m}_A = m_A \nu_A$ and $\hat{m}_B = m_B \nu_B$, and $U_o'(\mathbf{1}) = (U_{oA}'(\mathbf{1}), U_{oB}'(\mathbf{1}))$.

For Gaussian chains, the expression for the scattering function can be simplified further since then

$$\hat{\phi}_n = \phi^n \mathbf{1} M_{on} \nu_n$$

where ϕ is a diagonal matrix with the elements

$$\phi_{AA} = \exp(-b_{AA}^2 q^2/6); \quad \phi_{BB} = \exp(-b_{BB}^2 q^2/6) \tag{C.112}$$

and where b_{AA} and b_{BB} are the effective bond lengths between two A and two B units, respectively. Hence

$$U_o'(\mathbf{1}) = M_o\nu + \langle N(1) \rangle \ \phi[\mathbf{1} - \mathbf{P} \cdot \phi]^{-1} M_o\nu \tag{C.113}$$

where $M_o\nu$ is a column vector with the two components $M_{oA}\nu_A$ and $M_{oB}\nu_B$.

The Eq. (C.111) or (C.113) hold for copolymers where the functional groups of each kind of monomer have the same reactivity. However, they also hold for functional groups of unlike reactivity, and the only difference in the latter case lies within the transition-probability matrix $\mathbf{P}$. For unlike reactivities, the scalar elements p_{AA} and p_{BB} in the matrix for like functional groups, are now square matrices, and $\mathbf{p}_{AB}$ and $\mathbf{p}_{BA}$ are rectangular matrices. For instance, if the monomer A has two unlike functional groups, and monomer B four, then p_{AA} is a 2×2 and p_{BB} a 4×4 matrix, and $\mathbf{P}$ is of the rank 6×6. The rank of the population matrix in the first generation depends upon the order of the number of components, and is in the present case of the rank 2×2[95, 135].

2. Interdependences of Link Probabilities: Gordon's Basic Theorem[136]

The elements of the matrix $\langle N(1) \rangle$ are the various link probabilities which are not fully independent of each other. They have to fulfil the condition that the number of bonds formed by the A-monomers with B-monomers must equal the number of bonds formed by the B-monomers with A-monomers. The number of bonds which an A-monomer, selected at random, has formed with a B-monomer, is $\partial F_{oA}/\partial s_B$, and correspondingly

$\partial F_{oB}/\partial s_A$ is the number of links which a B-monomer, selected at random, has formed with A-monomers. Thus, $N_A \partial F_{oA}/\partial s_B$ is the total number of bonds formed by the A-units with a B-unit, and $N_B \partial F_{oB}/\partial s_A$ the total number of bonds formed by the B-units with an A unit, where N_A and N_B are the total number of A and B monomers in the system. Hence

$$n_A \partial F_{oA}/\partial s_B = n_B \partial F_{oB}/\partial s_A$$

or   (C.114)

$$n_A \langle (1) \rangle_{AB} = n_B \langle N(1) \rangle_{BA}$$

where n_A and n_B are the mole fractions of the two monomers in the polymer with $n_A + n_B = 1$. This basic relationship was first derived by Gordon[136]. (Note: the *mole* fractions are needed in Eq. (C.114), while in the formula for the scattering function always the mass fractions $m_A = n_A M_{oA}/(n_A M_{oA} + n_B M_{oB})$ and m_B are required).

3. The Number Average Molecular Weight M_n

Gordon's basic theorem Eq. (C.114) plays an important role since it provides a relationship between the link probabilities and the composition of the copolymer. The composition of a polymer can mostly be determined independently by spectroscopy or other techniques, and therefore, by the use of Gordon's theorem, one of the link probabilities can be expressed by the mole fractions n_A and n_B. For the determination of the other link probabilities, the weight-average molecular weight M_w can in principle be taken. However, because of the difference in the refractive indices of the various monomers, only an apparent molecular weight can be measured by light scattering. The number average molecular weight, on the other hand, can be measured correctly by osmometry if the molecules are not too large. The number average molecular weight can be calculated in every case by stoichiometric considerations. Gordon[28] has transformed a general stoichiometric formula given by Stockmayer[7b] into the generating-function formalism and obtained

$$M_n = \frac{(\mathbf{n} \cdot \mathbf{M}_0)}{1 - 0.5 \, (\mathbf{F}_0'(1) \cdot \mathbf{n})}$$ (C.115)

where $\mathbf{F}_0'(1)$ has the two components

$$F_{0A}'(1) = \partial F_{0A}/\partial s_A + \partial F_{0A}/\partial s_B = \langle N(1) \rangle_A$$
$$F_{0B}'(1) = \partial F_{0B}/\partial s_A + \partial F_{0B}/\partial s_B = \langle N(1) \rangle_B$$ (C.116)

4. Random Copoly-Condensates of A_2 with B_f Monomers[28, 137, 138]

The example of the random polycocondensation may be discussed in some detail. Let α be the reaction probability for a functional group in the A_2 monomer, and β the corresponding probability of a B-group in the B_f monomer. We have now to distinguish whether an A group has reacted with another A group or with a B-group of a f-functional B_f monomer. We denote the fraction of A-groups which have reacted with another A-group by $1 - p$ and consequently p is the fraction of A-groups which have reacted with a B-group; similar, we denote by ϱ the fraction of B-groups reacted with an A-group and $1 - \varrho$ the fraction of B-B bonds. The probability-generating functions for the various generations are then[139, 140]

$$
\begin{aligned}
F_{0A}(s) &= (1 - \alpha + \alpha p s_B + \alpha(1 - p) s_A)^2 \\
F_{0B}(s) &= (1 - \beta + \beta \varrho s_A + \beta(1 - \varrho) s_B)^f \\
F_{nA}(s) &= (1 - \alpha + \alpha p s_B + \alpha(1 - p) s_A) \\
F_{nB}(s) &= (1 - \beta + \beta \varrho s_A + \beta(1 - \varrho) s_B)^{f-1}
\end{aligned}
\qquad (C.117)
$$

Hence

$$
\langle \mathbf{N}(1) \rangle = \begin{bmatrix} \partial F_{0A}/\partial s_A & \partial F_{0A}/\partial s_B \\ \partial F_{0B}/\partial s_B & \partial F_{0B}/\partial s_B \end{bmatrix} = \begin{bmatrix} 2\alpha(1-p) & 2\alpha p \\ f\beta\varrho & f\beta(1-\varrho) \end{bmatrix} \qquad (C.118\,a)
$$

and

$$
\mathbf{P} = \begin{bmatrix} \partial F_{nA}/\partial s_A & \partial F_{nA}/\partial s_B \\ \partial F_{nB}/\partial s_A & \partial F_{nB}/\partial s_B \end{bmatrix} \qquad (C.118\,b)
$$

These matrices have to be inserted into Eq. (C.111) from which by the use of Eq. (B.22 a) or Eq. (B.22 b) the mean-square radius of gyration can be calculated, and from Eq. (B.56) to Eq. (B.58) the dynamic scattering behavior and the apparent diffusion coefficient.

For comparison with experiment and for numerical calculations, the values of the various link probabilities α, β, p and ϱ must be known. They can be determined as follows:

(i)   The overall extents of reactions of the A and B groups can be determined by titration or spectroscopy. (Polyester formation is typical, where A may be an OH-group and B an acid group).

(ii) Gordon's basic theorem (Eq. (C.114)) yields

$$
n_A 2\alpha p = (1 - n_A) f\beta\varrho
$$

from which one of the link probabilities can be eliminated, e.g. ϱ. The composition of the monomers in the polymer can be determined by common techniques, i.e. NMR, IR or UV-spectroscopy.

(iii) The last probability, p, can now be determined from either the number-average molecular weight or the apparent weight-average molecular weight, where M_{app} results from Eq. (C.113) by setting $\phi_n = 1$

$$
M_{app} = M_0\nu_0 + \langle \mathbf{N}(1) \rangle [1 - \mathbf{P}]^{-1} M_0\nu_0 \qquad (C.113\,a)
$$

with $M_0\nu_0 = (M_{0A}\nu_A, M_{0B}\nu_B)$

In practice, p is adjusted in the numerical calculations until the calculated M_{app} agrees with the measured one. All other quantities, for instance the molecular-weight dependence of $\langle S^2 \rangle_z$, or the shape of the scattering curve, and the whole dynamic scattering behaviour, are now fully determined if the effective bond length is known.

5. Copolymerization with Unlike Functional Groups

As an example, star-molecules may be considered, where linear chains have been grafted onto a branched nucleus[114]. The nucleus may be of the $A\!-\!\!<^B_C$ type, and the linear chains be made up of A-B bifunctional monomers. The nucleus of the mentioned type has only one end-group A per molecule but many free end-groups B, and the latter can be used for coupling the linear chains with their corresponding free A-end-groups. For the nucleus, only reactions between B and A or C and A are permitted, and for the linear chain only reactions between A and B. The extents of reaction may be α_A, α_B, and α_C for the functional groups A, B and C of the monomers which formed the nucleus, and β for the A and the B groups of the linear chain monomers. The fraction of free B-end-groups in the nucleus is then $(1 - \alpha_B)$. A certain fraction p of these free ends will have been used for the coupling reaction with linear chains. Conversely, the linear chains have a fraction $(1 - \beta)$ of free end-groups, and thus $(1 - \beta)\varrho$ may be the extent of coupling for the one (A) end-group of the chain. Since we do not wish to have free chains in the system, $\varrho = 1$ is assumed, i.e. all chains are fixed with their A-ends to the nucleus[141]. Altogether, we have the following list for the extent of reactions (Table 1):

Table 1. List of the various functional groups in the star-molecule with a branched nucleus, the corresponding extents of reaction, and the auxiliary variables denoting the partner of reaction

Functional	group	Extent of Reaction	Auxiliary variables
Blocks:			
Nucleus	A_N	α_A	s_B, s_C
	B_N	α_B	s_A
	C_N	α_C	s_A
Ray	A_R	β	s_{BR}
	B_R	β	s_{AR}
Coupling:			
Nucleus-Ray	$B_N \rightarrow A_R$	$(1 - \alpha_B)\,p$	s_{AR}
Ray-Nucleus	$A_R \rightarrow B_N$	$(1 - \beta)$	s_B

Star-molecules of this kind have been prepared by Pfannemüller[142-146] who used glycogen and amylopectin as nuclei. These branched polysaccharides have a large number of outer chains with more than 4 glucose units in length. Therefore, these outer chains with non-reducing end-groups could be used as primers (starting centres) for the enzyme phosphorylase. In the presence of glucose-1-phosphate, this enzyme extends the chains, the phosphate group in C1 position of the sugar is cleaved, and simultaneously the monomer is attached to the C4 group of an outer chain of the branched nucleus by the formation of an $\alpha\,(1 \to 4)$ glycosidic bond.

The mathematics of the star-structure formation was treated by Franken and Burchard and studied by light scattering techniques. Figure 14 shows a sketch of such star molecules, represented here as a so-called *directed* graph. The single arrow indicates the $\alpha\,(1 \to 4)$ glycosidic linkage and the double arrow the $\alpha\,(1 \to 6)$ glycosidic linkage, which is responsible for the extent of branching. Figure 14 a and b exhibit two typical rooted trees where in the one case a unit from the nucleus and in the other a unit from a ray was chosen.

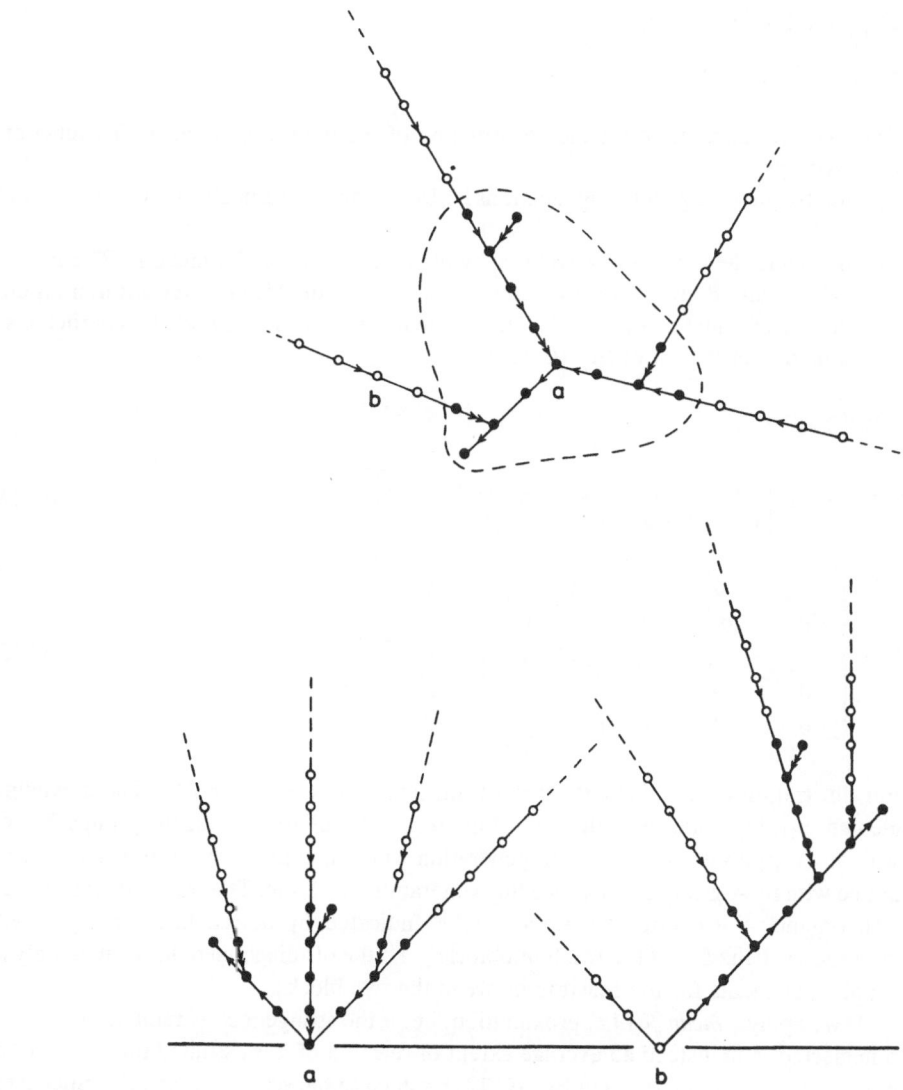

Fig. 14. Two rooted tree representatives of the molecular tree[114]

Following now the rules for setting up generating functions, one obtains the following set of p.g.f.s for the various generations:

$$F_{0N} = (1 - \alpha_A + \alpha_B s_B + \alpha_C s_C)(1 - \alpha_C + \alpha_C s_A)$$
$$\qquad\quad ((1 - \alpha_B)(1 - p) + (1 - \alpha_B) p s_{AR} + \alpha_B s_A)$$

$$F_{0R} = ((1 - \beta) s_B + \beta s_{BR})(1 - \beta + \beta s_{AR})$$

$$F_{nA} = ((1 - \alpha_B)(1 - p) + (1 - \alpha_B) p s_{AR} + \alpha_B s_A)(1 - \alpha_C + \alpha_C s_A)$$

$$F_{nB} = (1 - \alpha_A + \alpha_B s_B + \alpha_C s_C)(1 - \alpha_C + \alpha_C s_A) \qquad\qquad (C.119)$$

$$F_{nC} = ((1 - \alpha_B)(1 - p) + (1 - \alpha_B) p s_{AR} + \alpha_B s_A)(1 - \alpha_A + \alpha_B s_B + \alpha_C s_C)$$

$$F_{nAR} = ((1 - \beta) s_N + \beta s_{BR})$$

$$F_{nBR} = (1 - \beta + \beta s_{AR})$$

The cascade substitution for the construction of the path-weight generating function is achieved by
(i) multiplying the generating functions F_{nj} by $s^{\phi_{nN}}$ or $s^{\phi_{nR}}$, respectively, which yields U_{nj}, and
(ii) replacing the auxiliary variables s_j by the corresponding U_j functions. The matrices $\langle N(1) \rangle$ and P are then the coefficient matrices of the $U'_{nj}(1)$ after differentiation of the set of equations at $s = 1$. The elements are just the sum of the coefficients of similar s_j in the set of Eq. (C.118).

Thus[147],

$$\langle N(1) \rangle = \begin{bmatrix} \alpha_A & \alpha_B & \alpha_C & (1 - \alpha_B)p & 0 \\ 0 & 1 - \beta & 0 & \beta & \beta \end{bmatrix} \qquad\qquad (C.120)$$

$$P = \begin{bmatrix} \alpha_A & 0 & 0 & (1 - \alpha_B)p & 0 \\ \alpha_C & \alpha_B & \alpha_C & & 0 \\ \alpha_B & \alpha_B & \alpha_C & (1 - \alpha_B)p & 0 \\ 0 & 1 - \beta & 0 & 0 & \beta \\ 0 & 0 & 0 & \beta & 0 \end{bmatrix} \qquad\qquad (C.121)$$

For illustration of the elements in the transition matrix P see Fig. 15. The meaning of element p_{11}, for instance, is the sum of the reaction probabilities for the groups B and C with an A-group in the $(n + 1)$-th generation when the unit in the n-th generation was linked with its A-group to the preceding generation $n - 1$, etc. The transition matrix has a well organized structure: the square fields indicated by dotted lines in Eq. (C.121) describe the behavior of the two homoblocks, and the off-diagonal fields contain only the coupling elements for the reaction between the two blocks.

If we apply a *mean-field* approximation, i.e. if the stringent constraint $\alpha_B + \alpha_C = \alpha_A$ is neglected, and instead an average extent of reaction of α is assumed for all functional groups, the two submatrices in Eq. (C.121) reduce to the scalars 2α and β, respectively, and the matrix $\langle N(1) \rangle$ and P are now given by

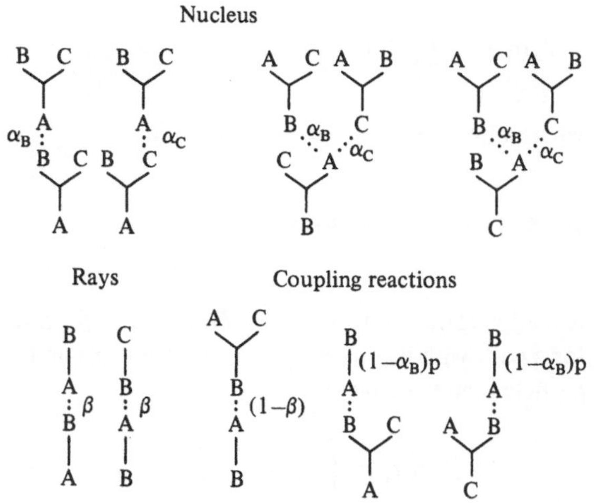

Nucleus

Rays   Coupling reactions

Fig. 15. The various transition probabilities for the star-copolymer of Fig. 14

$$\langle \mathbf{N}(1) \rangle = \begin{bmatrix} 3\alpha & 3(1-\alpha)p \\ 1-\beta & 2\beta \end{bmatrix} \tag{C.122}$$

$$\mathbf{p} = \begin{bmatrix} 2\alpha & 2(1-\alpha)p \\ 1-\beta & \beta \end{bmatrix} \tag{C.123}$$

which describe the properties of star-molecules with a nucleus obtained by random three-functional polycondensation (the factors 3 and 2 for some elements appear since in the random case in the zero-th generation we have three functional groups which could be involved in a coupling reaction; for all other generations the number of functional groups is two). All properties of the static and dynamic light scattering are now calculated from the general formulae for a general copolymer.

D. Properties of the Static and Dynamic Scattering Functions

I. Static Light Scattering

1. The Particle Scattering Function

In the derivation of the particle scattering factors we have reduced Debye's general formula to a sum over the average number of scattering elements $\langle N(n) \rangle$ (monomeric units) in the n-th generation (or shell of neighbours, from a unit selected at random) multiplied by a weighting function $\phi_n = \langle \exp(i\mathbf{q} \cdot \mathbf{r}_n) \rangle$. In general, this gives

$$P_w P_z(q^2) = \sum_{n=0}^{\infty} \langle N(n) \rangle \, \phi_n \tag{D.1}$$

where after averaging over all orientations

$$\phi_n = \int_0^{\infty} W(r_n) \, r_n \sin(q r_n) \, dr_n \tag{D.2}$$

Equation (D.2) shows ϕ_n as the *Fourier transform* of the pair distance distribution $W(r_n)$ for a path with its one end at $r = 0$ (the root) and the other at r_n (n-th generation). The particle scattering factor

$$P_z(q^2) = \frac{\sum \langle N(n) \rangle \int_0^{\infty} W(r_n) \, r_n \sin(q r_n) \, dr_n}{\sum \langle N(n) \rangle} \tag{D.1'}$$

is thus the average of these path-length Fourier transforms over all path-lengths in the molecule.

For the discussion of the properties of the static structure factors, it is often more convenient to write the scattering functions in terms of a space correlation function: $\gamma(r) \, 4 \pi r^2 \, dr$.

$$P_w P_z(q^2) = 4\pi \int_0^{\infty} \gamma(r) \, r \sin(qr) \, dr \tag{D.3}$$

The meaning of $\gamma(r)$ is as follows[149]: Select a unit at random and place it at the position $r = 0$. One may then ask for the probability of finding another unit at r in a volume element dr; $\gamma(r) \, dr$ is then the average of this probability over all units which were selected and positioned at $r = 0$. In other words, $\gamma(r) \, 4 \pi r^2 \, dr$ is the average number of units in a shell of thickness dr in a distance r from a unit selected at random. Integration over all shells must yield, therefore, the weight-average degree of polymerization

$$4\pi \int_0^{\infty} \gamma(r) \, r^2 \, dr = P_w \tag{D.4}$$

The space correlation function $\gamma(r)$ is closely related to the distribution of the units over the various shells $\langle N(n) \rangle$. The main difference consists of the fact that $4\pi \gamma(r) \, r^2 \, dr$ measures the number of units in a shell at *distance r* from a given unit while $\langle N(n) \rangle$ represents the number of *units in the n-th-shell*, where nothing needs to be known about the distance of the shell from the origin. This distance is taken into consideration by the extra factor $\phi_n(r)$ in Eq. (D.1'). Both functions, the space correlation function $\gamma(r)$ and the population number $\langle N(n) \rangle$, have their advantages and disadvantages. It should be mentioned, however, that $\langle N(n) \rangle$ is much easier and more directly calculated by application of the cascade procedure than the space correlation function, which can be obtained

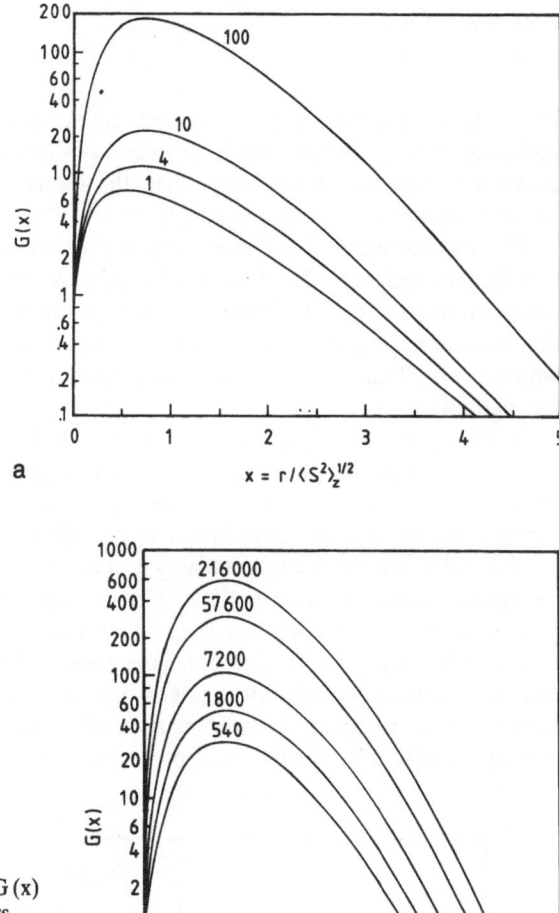

Fig. 16. a Space-correlation function $G(x)$ $= 4\pi\gamma(x)x^2$ of star-molecules with rays which obey the most probable length distribution[90, 102]. The numbers denote the numbers of rays per molecule which have in all cases the same average length. **b** Space-correlation function $G(x)$ of regular star-molecules with 18 rays of various length as indicated by the numbers[152]

by Fourier transforming the particle scattering factor and therefore requires knowledge of this function. We will make use of both expressions in the sequel. Figures 16 a and 16 b show two examples of space correlation functions, and Figure 22 displays the population $\langle N(n)\rangle$ for the three examples of a random polycondensate, an AB_2-polycondensate, and a monodisperse fraction of the random polycondensate, where all products have the same degree of polymerization.

The space correlation function has been extensively used by Kratky and Porod[148, 149] for monodisperse particles and was also introduced by Pekeris[150] and by Debye[151] for a description of spatial inhomogeneities in condensed matter. $\gamma(r)$ always has the properties

$$\gamma(r) = 1 \quad \text{for } r = dr$$
$$\gamma(r) = 0 \quad \text{for } r \to \infty \tag{D.5}$$

where dr is the radius of gyration of one monomeric unit. In fact, within the sphere or radius dr, one is always certain to find the selected unit. The other limit is fulfilled because at distances much larger than the radius of gyration there will be no other monomeric unit of the same molecule.

Clearly, the space correlation function $\gamma(r)$ is closely related to the segment-density distribution, and since the scattering intensity as a function of the scattering angle is the Fourier transform of $\gamma(r)$, there exists a close relationship between the density distribution and the angular distribution of the scattered light. This relationship often allows an immediate qualitative interpretation of scattering curves without time consuming model calculations.

In Fig. 17 to 19 the particle-scattering factors for some regularly branched and some polydisperse molecules are shown in plots of $P_z(q^2)^{-1}$ as function of $q^2 \langle S^2 \rangle_z$ (see also Table 2). The curves demonstrate clearly that branching causes an upturn while polydispersity tends to balance the influence of branching[34, 90].

The strong upturn due to branching is qualitatively well understood by comparing the corresponding density distributions. The strongest upturn is noticed with spheres of uniform density. Monodisperse linear chains on the other hand exhibit only a slight upturn at large $q^2 \langle S^2 \rangle$. In star-molecules several chains become affixed to a branching point, and thus the density of segments is increased. Moreover, as was first shown by Stockmayer and Fixman[91], the density profile is changed more and more towards the uniform density of a sphere. A similar, but even more pronounced approach of the

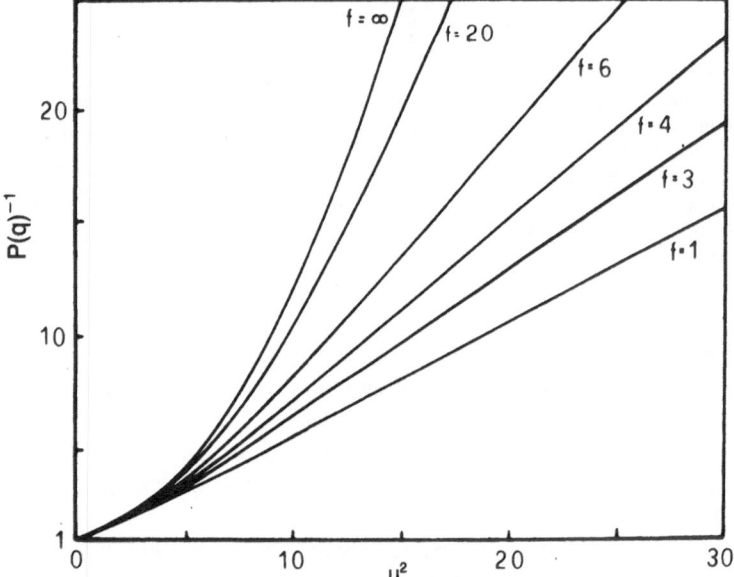

Fig. 17. Reciprocal particle-scattering factors of regular star-molecules (f = 1 represents the monodisperse linear chain)[88, 90]

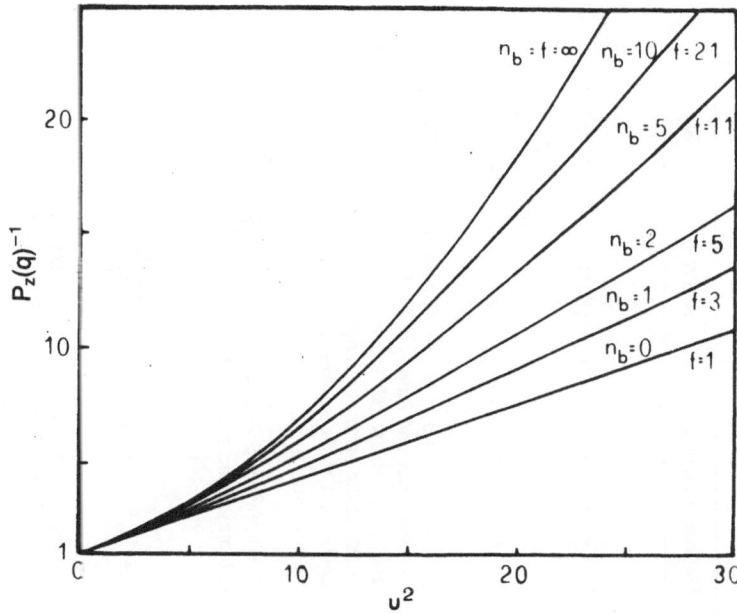

Fig. 18. Reciprocal particle-scattering factors of star-molecules with polydisperse rays, where f denotes the number of rays per molecule. The same functions are obtained also for the ABC-type polycondensates, where n_b denotes the number of branching points per molecule. The case $f = 1$ or $n_b = 0$ is identical to linear chains obeying the most probable lengths distribution. It also represents the scattering behaviour of randomly branched f-functional polycondensates[90]

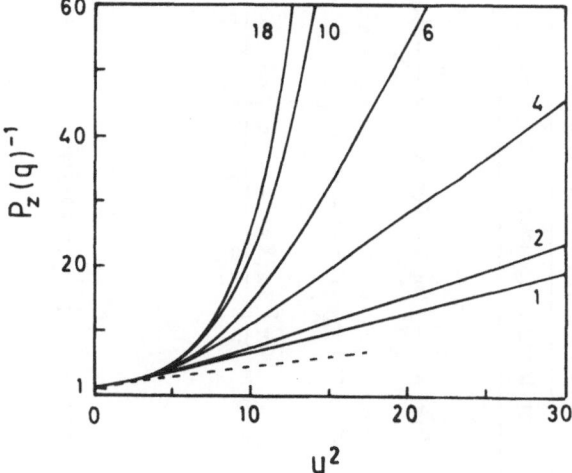

Fig. 19. Reciprocal particle-scattering factor of a tri-functional regularly branched molecule, "soft sphere model"[93], where the numbers denote the shells of branching (details of the model are shown in Fig. 24 a, $u^2 = \langle S^2 \rangle q^2$)

sphere density has been predicted by Daoud and Cotton[154] for star-molecules with strong thermodynamic interaction amongst the monomeric units.

Certainly, the approach to the "hard sphere density" will be closer for the regularly branched three-functional model. A glance at Fig. 19 reveals a behavior of $P(q^2)^{-1}$ which is fully consistent with this picture.

Table 2. Particle scattering factors of some selected models

Model	Particle scattering factor	Defining parameters	Autor	Ref.
A. Rigid and regular shapes				
Sphere	$P(\theta) = \left[\dfrac{3}{X^3}(\sin X - X\cos X)\right]^2 = \dfrac{9\pi}{2X^3}\,J_{3/2}^2(X)$	X = rq, r = radius	(Rayleigh, 1914)	233
Hollow sphere, infinitely thin	$P(\theta) = \left(\dfrac{1}{2X^2}\right)(1 - \cos 2X) = \left(\dfrac{\sin X}{X}\right)^2$	X = rq, r = radius	(Tinker, 1972; Oster and Riley, 1952)	234 235
Hollow sphere, finite shell thickness	$P(\theta) = \dfrac{9\pi}{2}\left[\dfrac{J_{3/2}(X_0)}{X_0^{3/2}} - \dfrac{n_0-n_i}{n_0-n_s}\left(\dfrac{r_i}{r_0}\right)^3\dfrac{J_{3/2}(X_i)}{X_i^{3/2}}\right]^2$	$X_0 = r_0 q$, $X_i = r_i q$; r_0, r_i are outer and inner radii, n_0, n_i, n_s are refractive indices of the outer and inner region, and of the surrounding medium	(Kerker et al., 1962)	236
Ellipsoid of revolution	$P(\theta) = \dfrac{9\pi}{2}\displaystyle\int_0^{\pi/2} \dfrac{J^2(V)}{V^3}\cos\beta\,d\beta$	$V^2 = q^2(a^2\cos^2\beta + b^2\sin^2\beta)$; a, b major and minor semi-naxes of the ellipsoid	(Porod, 1948)	149
Rod, infinitely thin	$P(\theta) = \dfrac{2}{X}\left[\displaystyle\int_0^x \dfrac{\sin t}{t}\,dt\right] - \left(\dfrac{\sin X/2}{X/2}\right)^2$	X = lq, l = rod length	Neugebauer, 1943	237
Disc, infinitely thin	$P(\theta) = \dfrac{2}{X^2}[1 - J_1(2X)/X]$	X = rq, r = radius, $J_1(X)$ = Bessel Function of the 1st order	(Kratky and Porod, 1949)	238
Circular cylinder	$P(\theta) = \displaystyle\int_0^{\pi/2} \dfrac{\pi}{X_l\cos\beta}\left[J\left(\dfrac{X_l\cos\beta}{2}\right)\right] \times \left[\dfrac{2J_1(X_r\sin\beta)}{X_r\sin\beta}\right]^2 \sin\beta\,d\beta$	$X_l = lq$, $X_r = rq$; l = cylinder length, r = radius of the cylinder	(Fournet, 1951; Mittelbach and Porod, 1961)	239 240

Note:

$$P(\theta)_{cylinder} = \begin{cases} P(\theta)_{Rod} & \text{for } r \to 0 \\ P(\theta)_{Disc} & \text{for } l \to 0 \end{cases}$$

B. Linear chain molecules (long Gaussian chains)

Coil, monodisperse (Schulz, 1935; Flory, 1936 distribution)

$$P(\theta) = \frac{2}{U^4}[\exp(-U^2) - (1 - U^2)]$$

$$P_z(\theta) = [1 + U^2/3]^{-1}$$

$U^2 = \langle S^2 \rangle q^2 \langle S^2 \rangle$
mean square radius of gyration — (Debye, 1945, 1964) 241

$U^2 = \langle S^2 \rangle_z q^2$; the index indicates the z-average — Zimm, 1948) 12

Coil, polydisperse m head-to-tail coupled chains (Schulz, 1939[123]; Zimm, 1948 distribution)[12]

$$P_z(\theta) = \frac{1}{m+1}\left\{\frac{2}{m}\sum_{j=1}^{m}(m+1-j)\left[1 + \frac{U^2}{m+2}\right]^j\right\}$$

$U^2 = \langle S^2 \rangle_z q^2$ m = $[(DP_w/DP_n) - 1]^{-1}$ DP_w, DP_n weight and number average degree of polymerization respectively — (Franken and Burchard, 1973); 200

C. Branched chain molecules (Gaussian subchains)

Regular Star,

$$\left[P(\theta) = \frac{2}{fV^2}\left[V - (1 - \exp(-V)) + \frac{f-1}{2}(1 - \exp(-V))^2\right]\right]$$

$V = \dfrac{f}{3f-2}U^2$

$U^2 = \langle S^2 \rangle_z q^2$; f number of rays — $\begin{cases}(\text{Benoit, 1953; } 89\\ \text{Burchard, 1974)}88\end{cases}$

Star, polydisperse, Schulz-Flory distribution $P_z(\theta) =$

$$\frac{1 + U^2/3f}{[1 + U^2(f+1)/6f]^2}$$

$U^2 = \langle S^2 \rangle_z q^2$; f number of rays — (Burchard, 1974 a, 1977) 88, 90

Soft sphere model

$$P(\theta) = N^{-2}[1 + 6P_1P_2 + 6X(P_0 + (P_1P_2)^2) + 3P_2^2P_6(3x-P_2 + (X+1)(P_2P_5 - P_3P_4))]$$

m = monomer units between branching points;
n = number of branching shells
$X = 2^n - 1$
N = degree of polymerization
b = effective bond length
— Burchard und Kajiwara (1982) 93

with

$P_0 = (1/y^4)[my^2 - (1 - \phi^m)]$

$P_2 = [1 - (2\phi^m)^n]/(1 - 2\phi^m)$

$P_4 = (1 - \phi^{2mn})/(1 - 2\phi^{2m})$

$P_6 = 1/(1 - 2\phi^m)$

$\phi = \exp(-b^2q^2/6)$

$P_1 = \phi(1 - \phi^m)/(1 - \phi)$

$P_3 = (1 - \phi^{2mn})/(1 - \phi^m)$

$P_5 = \phi^{m(n+1)}/(1 - 2\phi^{2m})$

Table 2 (continued)

Model	Particle scattering factor	Defining parameters	Autor	Ref.
Regular comb, monodisperse	$$P(\theta) = \frac{2}{X^4}\{X^2 - [1-\exp(-kX^2)] + [1-\exp(-X^2(1-k)/f)]$$ $$\times\left[f + 2\frac{1-\exp(-X^2kf/(f+1)}{1-\exp(-X^2k/(f+1)}\right] + [1-\exp(-X^2(1-k)/f)]^2$$ $$\times \frac{(f-1)[\exp(-X^2k/(f+1)-1] - [1-\exp(-X^2k(f-1)/(f+1)]}{[1-\exp(-X^2k/(f+1)]^2}\}$$	$X^2 = q^2(Nb^2/6)$, $k = N_0/N$ N total number of chain elements N chain elements of the back-bone f number of side branches, equally spaced along the back-bone	(Casassa and Berry, 1966)	92
Comb, monodisperse back-bone and monodisperse side chains, random coupling of the side chains along the back-bone	$$P_z(\theta) = \frac{2/X^4}{1+(1-k)^2/f}\left\{\begin{array}{l} X^2 - [1-\exp(-X^2k)] \\ + [1-\exp(-X^2(1-k)/f] \end{array}\right.$$ $$\times\left[f - 2\frac{1-\exp(-X^2k)}{X^2k/f}\right]$$ $$\left. + [1-\exp(-X^2(1-k)/f]^2\ \frac{X^2k - [1-\exp(-X^2k)]}{(X^2k/f)^2}\right\}$$	$X^2 = q^2\bar{N}_n$; $k = N_0/\bar{N}_n$; $\bar{N}_n$ number average of the total number of chain elements; f number average number of side chains, randomly distributed along the back-bone; N_0 chain elements of the back-bone.	(Casassa and Berry, 1966)	92
Randomly branched f-functional poly-condensate	$$P_z(\theta) = |1 + U^2/3|^{-1}$$	$U^2 = q^2\langle S^2\rangle_z$	(Kajiwara et al., 1970)	34
Polycondensate of the A—B type (B\B)	$$P_z(\theta) = [1 + U^2/6]^{-2}$$	$U^2 = q^2\langle S^2\rangle_z$	(Burchard, 1972, 1977)	90
B Type C Polycondensates A (B\C)	$$P_z(q) = \frac{1 + C\langle S^2\rangle_z q^2/3}{[1+(1+C)\langle S^2\rangle_z q^2/6]^2}$$	$C = (\beta^2 + \gamma^2)$ $$\cdot\left[(\beta+\gamma)+\frac{2\beta\gamma}{1-\beta-\gamma}\right]^{-1}$$	Burchard 1977	90

randomly cross-linked chains	$P_z(q) = P_{zp}(q)\dfrac{P_{wp}}{P_w}\dfrac{1}{1 - (P_{wp}P_{zp}(q)-1)\,\varepsilon f(m-1)} \approx \dfrac{(1-\varepsilon/\varepsilon_c)P_{zp}(q)}{1-(\varepsilon/\varepsilon_c)P_{zp}(q)}$	$P_w = P_{wp}\dfrac{(1-(P_w-1))^{-1}}{\varepsilon f(m-1)} = P_{wp}(1-\varepsilon/\varepsilon_c)^{-1}$	Whitney, Burchard (1980)	117
randomly cross-linked systems for common primary chain distribution	$P_wP_z[\vartheta(\phi)] = (1+\alpha\phi)f_w(\phi)/\{1-[f_w(\phi)-1]\alpha\phi\}$		Kajiwara and Gordon 1973	170

Name of distribution	$f_w(\phi)$	
1 Monodisperse	$[(1+\phi)/(1-\phi)] - [2\phi/(1-\phi)^2 y][1-\phi^y]$	
2 Poisson	$[(1+\phi)/(1-\phi)] - [2\phi/(1-\phi)^2 y_n]\{1-\exp[(-y_n(1-\phi)]\}$	y_n and y_w are number and weight average degrees of polymerization of primary chains
3 Random Schulz-Flory; bifunctional random condensation or radical disproportionation	$[y_n + \phi(y_n-1)]/[y_n - \phi(y_n-1)]$	$\phi = \exp(-b^2q^2/6)$
4 Self-convoluted random; radical combination	$[\pm y_n^2 + 2\phi y_n - \phi^2(y_n-2)^2]/[y_n - \phi(y_n-2)]^2$	
5 Multiple self-convoluted random, Schulz-Zimm	$[(1+\phi)/(1-\phi)] - [2\phi/(1-\phi)^2 y_n]\{1-[1-(y_w-y_n)\ln\phi]^{-y_n/(y_w-y_n)}\}$	

The diminishing effect of the upturn due to polydispersity[155] may be demonstrated with an example. Let us consider a mixture of two linear chains with molecular weights M_1 and M_2 and particle-scattering factors $P_1(q^2)$ and $P_2(q^2)$, respectively. The total particle-scattering factor is then given by

$$P_z(q^2) = [w_1 M_1 P_1(q^2) + w_2 M_2 P_2(q^2)]/M_w \qquad (D.6)$$

where w_1 and w_2 are the weight fractions of the two polymers. Let each of the two components follow a linear relationship for $P(q^2)^{-1}$ as a function of q^2, i.e.

$$P_1(q^2) = (1 + 1/3 \langle S^2 \rangle_1 q^2)^{-1}$$
$$P_2(q^2) = (1 + 1/3 \langle S^2 \rangle_2 q^2)^{-1}$$

Let $\langle S^2 \rangle_2 = q \langle S^2 \rangle_1$ and $w_1 M_1 / M_w = w_2 M_2 / M_w = 1/2$; $P_z(q^2)$ can then be calculated. Figure 20 shows the curves for the two individual components and the sum. One notices that $P_2(q^2)$ has decayed to a very low value at a certain angle, say 50 °C while the second component still scatters the light strongly at even larger angles. Thus, at large angles the scattering behavior becomes more and more determined by the low molecular weight component which in the reciprocal plot has a straight line. At small angles on the other hand, both components contribute almost equally and this causes a steeper initial part in the reciprocal plot (see Fig. 20 below).

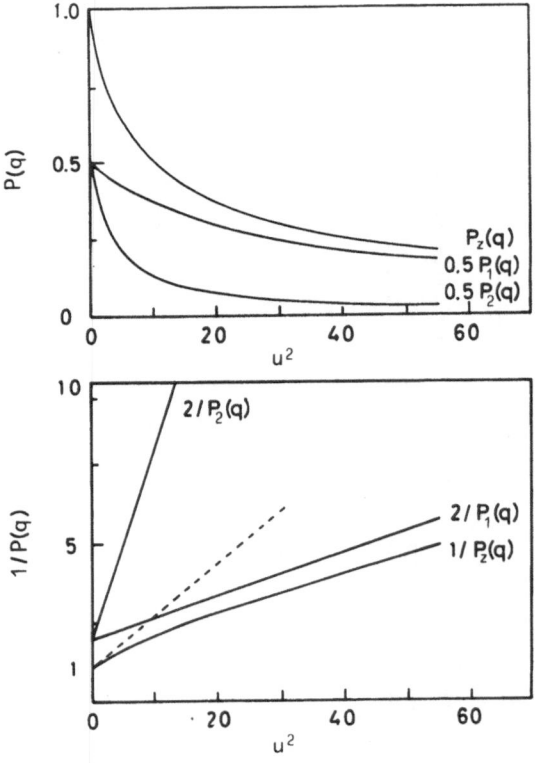

Fig. 20. *Above:* Particle-scattering factors $P_1(q)$ and $P_2(q)$ for two linear chains, where $\langle S^2 \rangle_2 = 9 \langle S^2 \rangle_1$, and the particle-scattering factor of a mixture of both chains with $w_1 M_1 = w_2 M_2$. *Below:* The corresponding reciprocal particle-scattering factors. Note the downturn for $P_z(q)$ at low q^2. The *dotted line* indicates the initial slope defined by $\langle S^2 \rangle_z$ of the mixture

For other types of particle-scattering factors, the effect of polydispersity will be similar. The "soft sphere model", for instance, shows a strong upturn of the reciprocal particle-scattering factor; a broad molecular-weight distribution will partially compensate this upturn. For randomly branched chains, the upturn becomes fully balanced by the huge width of the molecular-weight distribution of these samples to give a perfect straight line[34].

The same occurs with the slight upturn of linear chains when the chains have a most probable distribution[156]. In fact, the average number of units in the n-th generation decays in both cases in a similar manner like[34]

$$\langle N(n) \rangle = af[a(f-1)]^{n-1} \tag{D.7}$$

where $f = 2$ represents the linear polydisperse and $f > 2$ the randomly branched poly-condensate. The same type of distribution for $\langle N(n) \rangle$ is also obtained with randomly cross-linked linear chains if the latter obey the most probable distribution in length.

This leads to the conclusion that polydisperse linear chains cannot be distinguished from randomly branched chains using only the shape of their scattering curves. Indeed, when the link probabilities are expressed in terms of the mean-square radius of gyration, the particle-scattering factor is given in both cases by

$$P_z(q^2) = (1 + 1/3 \langle S^2 \rangle_z q^2)^{-1} \tag{D.8}$$

The straight line of $P_z(q^2)^{-1}$ as function of q^2 has in fact been observed for a number of non-fractionated branched materials[117, 137, 138, 157]. One example is depicted in Fig. 21.

The complete balance of the upturn by the polydispersity is only obtained for random branching processes. Often the reaction is impeded by serious constraints, or the primary chains before cross-linking are monodisperse. Then the resultant final molecular-weight distribution is narrower than in the random case, and the characteristic upturn as a result of branching, develops again. A strange coincidence in behavior is observed with star-molecules, where the rays are polydisperse, and with the ABC-type polycondensates. In both cases the particle-scattering factors can be expressed as[90]

$$P_z(q^2) = \frac{1 + C \langle S^2 \rangle_z q^2/3}{(1 + (1 + C) \langle S^2 \rangle_z q^2/6)^2} \tag{D.9}$$

where

$$C = 1/f \quad \text{for polydisperse rays in a star-molecule}$$

$$C = \left(1 + \frac{2a^2 p(1-p)}{1-a}\right)^{-1} \text{for the ABC-polycondensate}$$

The reason for this coincidence is discovered when the average number of units $\langle N(n) \rangle$ in the n-th generation is calculated. In both cases we find the same type of distribution, i.e.

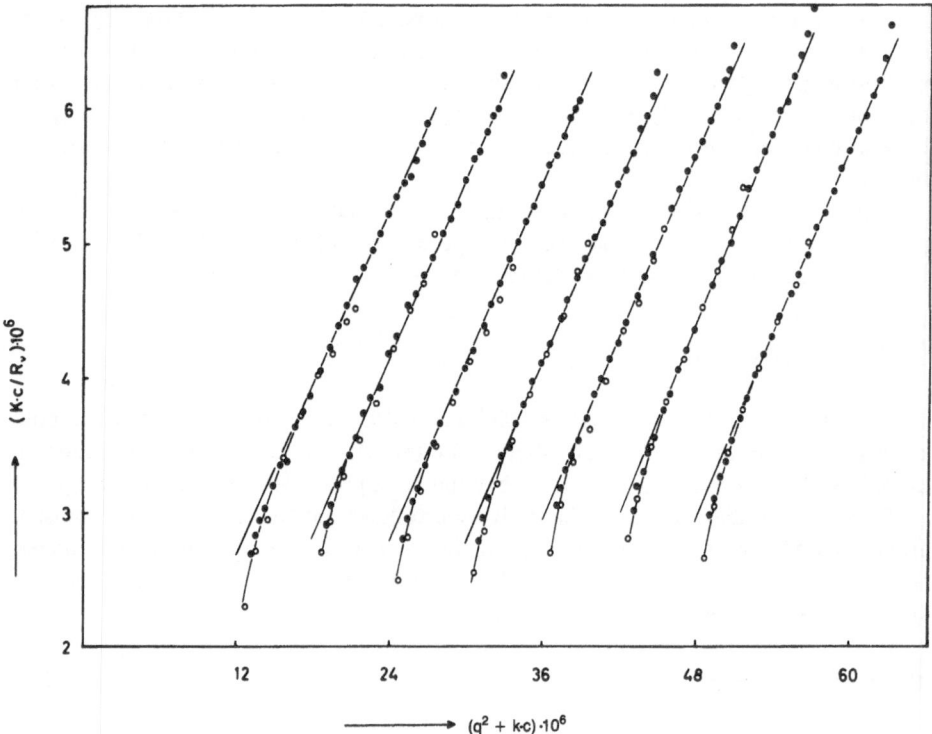

Fig. 21. Zimm-plot of an A_3/B_2 random polycondensate. A_3: benzene 1.3.5.-triacetic acid; B_2: decamethylene glycol. Measurement have been made in benzene at 20 °C, with the two wave lengths of $\lambda_0 = 546$ nm (●) and $\lambda_0 = 436$ nm (○). h = $(4\,\pi/\lambda)$ sin $\theta/2$[138]

$$\langle N(n) \rangle = 2\alpha^n + 2(n-1)\alpha^2 p(1-p)\alpha^{n-2} \qquad (D.10)$$
$$(\text{ABC-polycondensate})$$

$$\langle N(n) \rangle = 2\alpha^n + (n-1)(f-1)(1-\alpha)\alpha^{n-2} \qquad (D.11)$$
$$(\text{polydisperse stars})$$

which in fact shows the equivalence of f and $1 + 2\alpha^2 p(1-p)/(1-\alpha)$.

We already mentioned that it is the width of the molecular-weight distribution which smoothes out the upturn. In fact, the polydispersity increases like[7, 104]

$$M_w/M_n \sim M_w \qquad (D.12)$$

for the A_f polycondensate but only like[104]

$$M_w/M_n \sim M_w^{1/2} \qquad (D.13)$$

for the ABC-polycondensate. A real proof of this effect of polydispersity was, moreover, given by Kajiwara[83]. By application of the *Lagrange expansion*[132, 133] *technique* to a modified path-weight generating function (see Appendix), he derived a formula for the particle-scattering factor of monodisperse fractions from the A_f polycondensate. This fraction contains all isomers with different branching density but the same degree of polymerization x. The particle-scattering factor is given by

$$P_x(q^2) = 1 + \frac{x!}{x[(f-1)x]!} \sum_{n=1}^{x-1} [(f-1)\phi]^n \frac{[(f-1)x-n]!}{(x-n-1)!} \frac{[n(f-2)+f]}{(f-1)} \qquad (D.14)$$

Figure 23 shows as examples the reciprocal particle-scattering factors for $f = 2, 3$ and 6. Again, the upturn with increasing branching develops. One realizes from Eq. (D.14) that $P_x(q^2)$ becomes independent of the number of functional groups per monomer if f is large.

It is of interest to compare the average number of units in the n-th generation $\langle N(n) \rangle_x$ for these fractions of isomers with that of the polydisperse random polyconden-sate. This number was first calculated by Kurata and Fukatsu[158] and can again be obtained by Lagrange expansion[83] to give

$$\langle N(n) \rangle_x = \frac{x![(f-1)x-n]!}{[(f-1)x]!(x-n-1)!} (f-1)^{n-1} [n(f-2)+f] \qquad (D.15)$$

$$(n > 0)$$

For large x the factorials can be replaced by Stirling's approximation which simplifies Eq. (D.15) to

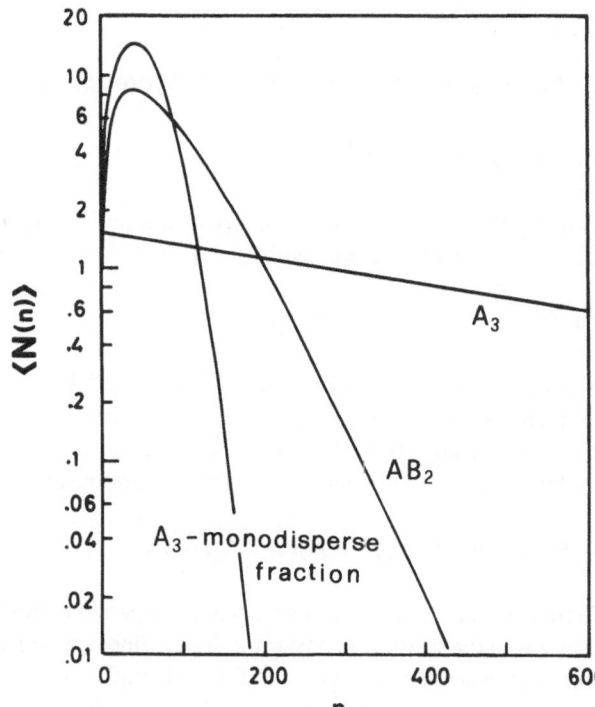

Fig. 22. Average number of un-its in the n-th generation for a three-functional *random* poly-condensate, A_3, the three-func-tional *restricted* polycondensate, AB_2, and a monodisperse frac-tion from A_3. $X_w = 1000$ in all cases

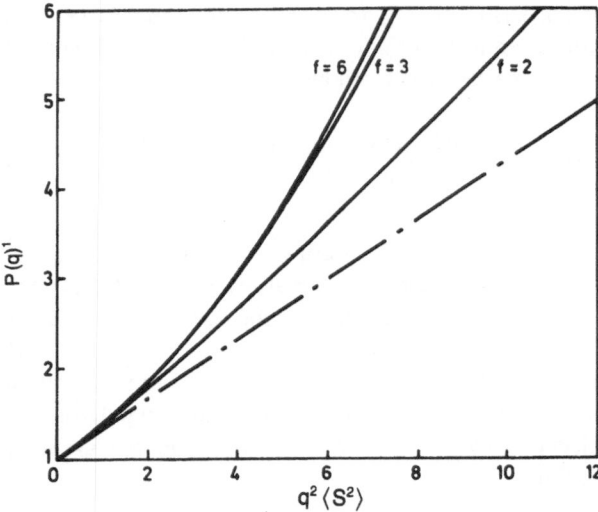

Fig. 23. Reciprocal particle-scattering factors of monodisperse randomly branched polycondensates of functionalities $f = 2, 3, 6$. The chain curve represents the polydisperse random polycondensate[83]

$$\ln \langle N(n) \rangle_x = ((f-1)x - n) \ln [1 - n/(f-1)x] - (x-n) \ln [1 - n/x]$$
$$+ \ln [n(f-2)/(f-1+f/(f-1))] \tag{D.15'}$$

Fig. 22 shows this function for $f = 3$ in comparison to the unfractionated A_3 and the AB_2 polycondensates. (A list of particle scattering factors for various structures is given in Table 2).

2. Various Plots for the Particle Scattering Factor

a) The Zimm-Plot[159]

In every case the first two members in the power expansion of the particle-scattering factor in terms of q^2 are given by

$$P_z(q^2)^{-1} = 1 + 1/3 \langle S^2 \rangle_z q^2 + 0(q^4) \tag{D.16}$$

Hence, at sufficiently small values of $u^2 = \langle S^2 \rangle_z q^2$ the mean-square radius of gyration can be evaluated from the slope of the initially linear part of $P_z(q^2)^{-1}$. This property led Zimm to the suggestion of plotting Kc/R_θ against $q^2 + kc$. According to Debye[160], the scattering intensity of solutions at a finite concentration is given by

$$Kc/R_\theta = 1/(M_w P_z(q^2)) + 2A_2c + 3A_3c^2 + \dots \tag{D.17}$$

Therefore, the Zimm-plot gives a family of parallel lines which appear shifted along the abscissa by the arbitrary constant k. These line can be extrapolated for each concentration to zero scattering angle, and for each angle an extrapolation to zero concentration can be carried out. This Zimm-plot finally results into the two limiting lines $Kc/R_\theta|_{c=0}$

and $Kc/R_\theta|_{\theta=0}$; the first yields $(M_w P_z(q^2)^{-1})$ from which the mean-square radius of gyration can be determined, and the latter gives $1/M_{app} = 1/M_w + 2A_2c + 3A_3c^2 \ldots$, where A_2, A_3, ... are the *virial* coefficients. Figure 21 shows an example of a Zimm-plot with linear angular dependence.

b) The Berry-Plot[161]

In many cases the upturn due to branching already starts at rather low q-values, and an extrapolation to zero angle can become very difficult in the common linear Zimm-plot. In these cases, it is often useful to apply the so-called *Berry-plot*[161], where the square root of Kc/R_q is plotted against $q^2 + kc$. This plot was suggested by Berry for polymers in good solvent, where the concentration dependence is no longer linear but is also affected by the third virial coefficient. To a large extent the third virial coefficient is determined by the second virial coefficient; for hard spheres statistical thermodynamics shows[162, 165]

$$A_3M = 5/8 (A_2M)^2 \tag{D.18}$$

Stockmayer and Casassa[163] realized that for flexible coils this factor varies with the coil expansion but still the more general equation

$$A_3M = g (A_2M)^2 \tag{D.18'}$$

can be used as a good approximation. If $g = 4/3$ (see Yamakawa)[164, 165], then $(1 + 2A_2Mc + 3A_3Mc^2)$ becomes $(1 + A_2Mc)^2$, and the light-scattering equation may be written

$$Kc/R_q = [1/(M_w P_z(q)] (1 + A_2M_wc)^2 \tag{D.19}$$

which would give a straight line if $(Kc/R_q)^{1/2}$ is plotted. For star-molecules it turned out that also the angular dependence becomes straightened out. This effect was first observed empirically. A look at Eq. (D.9) reveals that such behavior is expected also by theory[90].

c) The Guinier-Plot[166, 167]

Figure 24 shows a Berry-plot for the three-functional regularly branched chain model[93]. It shows that only for n = 1 (star-molecules) and n = 2 does this plot give a fairly good straight line; for larger shells of branching a much stronger upturn occurs which can no longer approximated by quadratic behavior. At such high branching we can expect that the density profile of the segments has approached that of a sphere with the two essential differences (i) the outer chains are dangling around and (ii) the chains are flexible and not rigid. We thus may call this model a "soft sphere"[93]. Many years ago, Guinier[166, 167] showed that the particle-scattering factor of globular structures can over a wide range of q^2 be approximated by the following equation

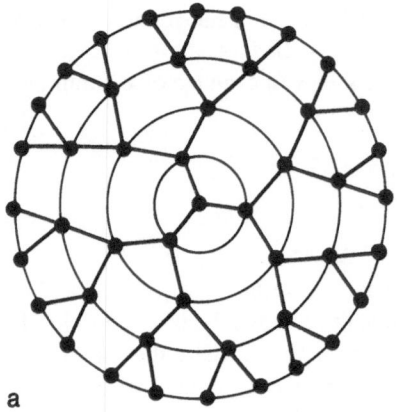

a

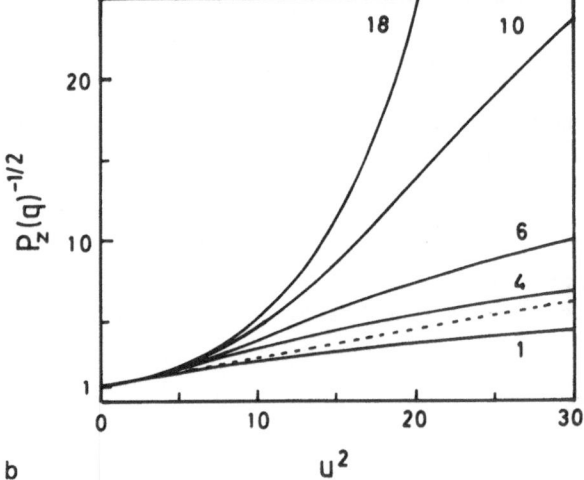

b

Fig. 24. a The three-functional regularly branched chain model with Gaussian behavior of the subchains, called the "soft sphere" model[93]. b The Berry-plot of the reciprocal particle-scattering factor of the "soft sphere" model. Compare also Figs. 19, 25 and 27

$$P_z(q^2)^{-1} = \exp(1/3\, q^2\, \langle S^2 \rangle_z) \qquad\qquad (D.20)$$

Thus, the *Guinier plot*, where $\ln P_z(q^2)^{-1}$ is plotted against q^2, should give a straight line for such globular structures. Figure 25 shows that for the "soft sphere" model this limit is indeed approached more and more closely with increasing branching shells. For comparison, the particle-scattering factor of the "hard sphere" is also plotted in Fig. 25. The points represent measurements from polyvinyl acetate microgels in methanol[168].

d) The Kratky Plot[169]

Finally we mention the *Kratky plot* which also may help to detect branching. Here $(\langle S^2 \rangle\, q^2)\, P_z(q^2)$ is plotted against q. Figures 26 and 27 show the Kratky plots for regular star-molecules and for the "soft sphere" model. Linear randomly coiled chains result in

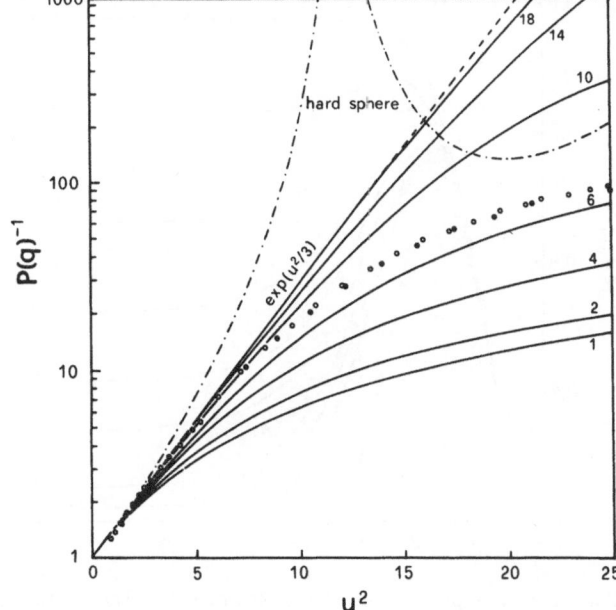

Fig. 25. Guinier plot of the soft sphere model. The numbers denote the number of branching shells; the filled and open circles are light-scattering results from polyvinyl acetate (PVAc) microgels in methanol at 20 °C at $\lambda_0 = 546$ nm and 436 nm, respectively. The dot-dash line corresponds to the Rayleigh-Gans behavior of hard spheres, i.e. no Mie scattering[93]

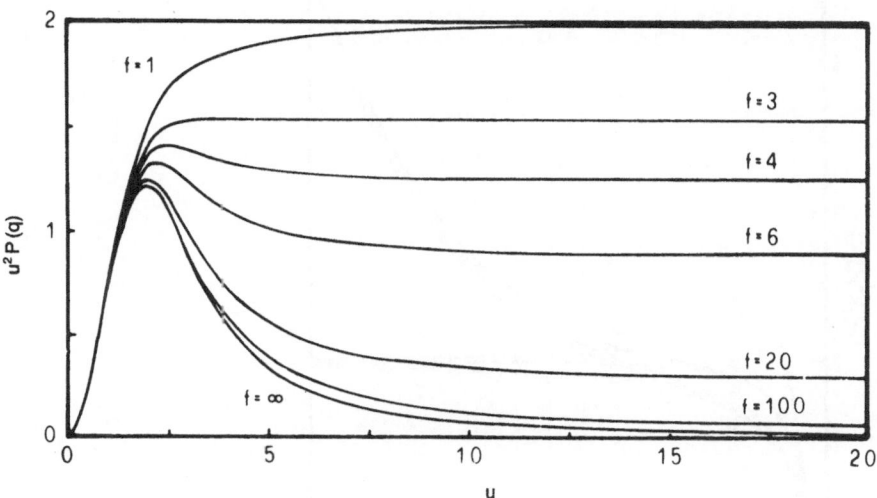

Fig. 26. Kratky plot for regular star-molecules[90]

an angular independent asymptote which has a value of 2 for monodisperse coils and 3 for polydisperse chains if $M_w/M_n = 2$[90]. The striking feature with branched chains is the appearance of a maximum for stars and other regularly branched chains[90]. This maximum becomes more and more pronounced with increasing branching density. Random f-functional polycondensates exhibit again the same behavior as polydisperse linear chains with $M_w/M_n = 2$. For randomly cross-linked chains such behavior is also obtained,

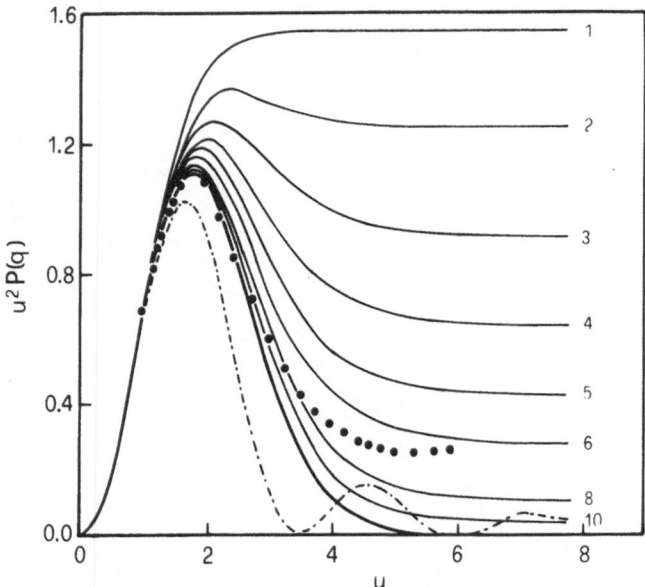

Fig. 27. Kratky plot for the soft sphere model[93]. The points represent light-scattering measurements form a PVAc microgel (compare also Figs. 19, 24 and 25)

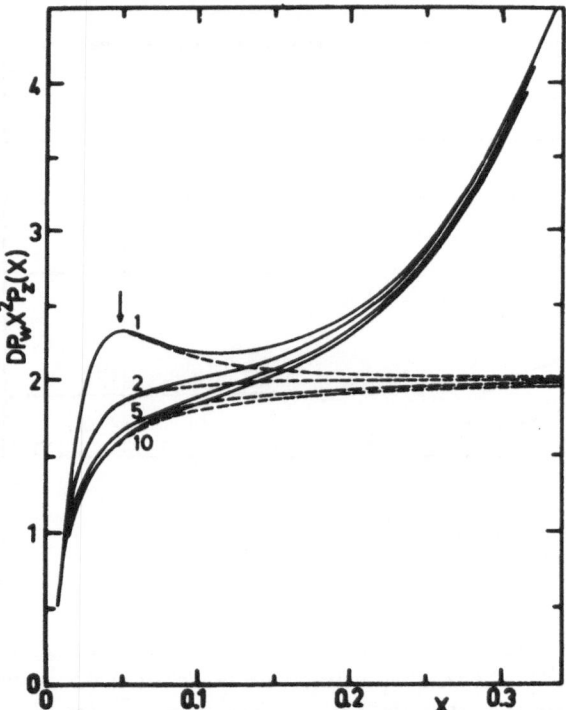

Fig. 28. Kratky plot for vulcanisates with different primary chain distributions. *Solid curves* for a/b = 6 (a = persistence length, b = bond length) and *dashed curves* for a/b = 1 (Gaussian chains). The numbers in the figure denote the value of P_{wp}/P_{np}, i.e. the polydispersity of the primary chains[119] $X^2 = q^2 \langle S^2 \rangle_z$

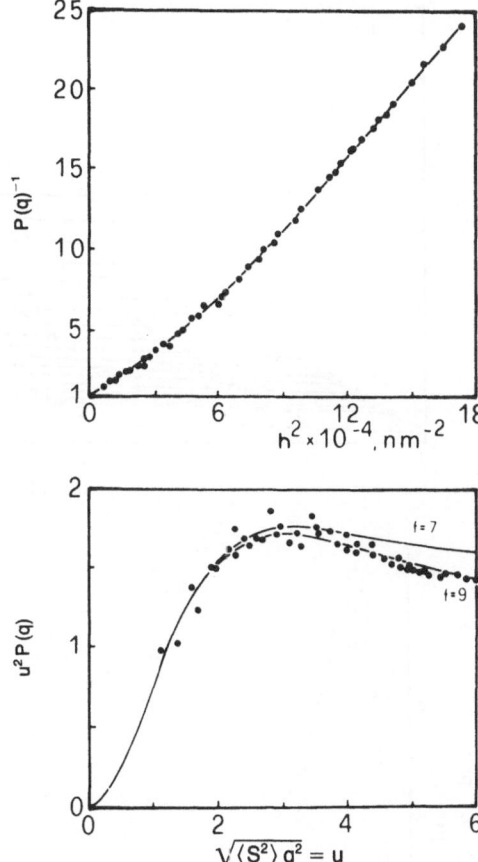

Fig. 29. Zimm-plot and Kratky-plot from a star-branched macromolecule. Star centre: nodule of densely cross-linked polydivinyl benzene; rays: monodisperse polystyrene chains. Measurement in toluene at 20 °C and with the three wavelengths $\lambda_0 = 546$, 436 and 365 nm[90]

but here only, if the primary chains before cross-linking obey the most probable length distribution. If the primary chains are less broad in their length distribution, the characteristic maximum shows up again[119, 170] (see Fig. 28). Kajiwara[84] has recently extended his calculation and derived equations for the scattering properties of monodisperse fractions of these cross-linked materials. As could be expected, the maximum becomes more and more pronounced with the increasing number of cross-linked chains. Two further examples of measurements are given in Figs. 29 and 30.

3. The Mean-Square Radius of Gyration

The first direct calculations for randomly branched polycondensates were carried out by Zimm and Stockmayer[13] by direct evaluation of the double sum

$$\langle S^2 \rangle = x^{-2} \sum_{j}^{x} \sum_{k}^{x} \langle r_{jk}^2 \rangle \tag{D.21}$$

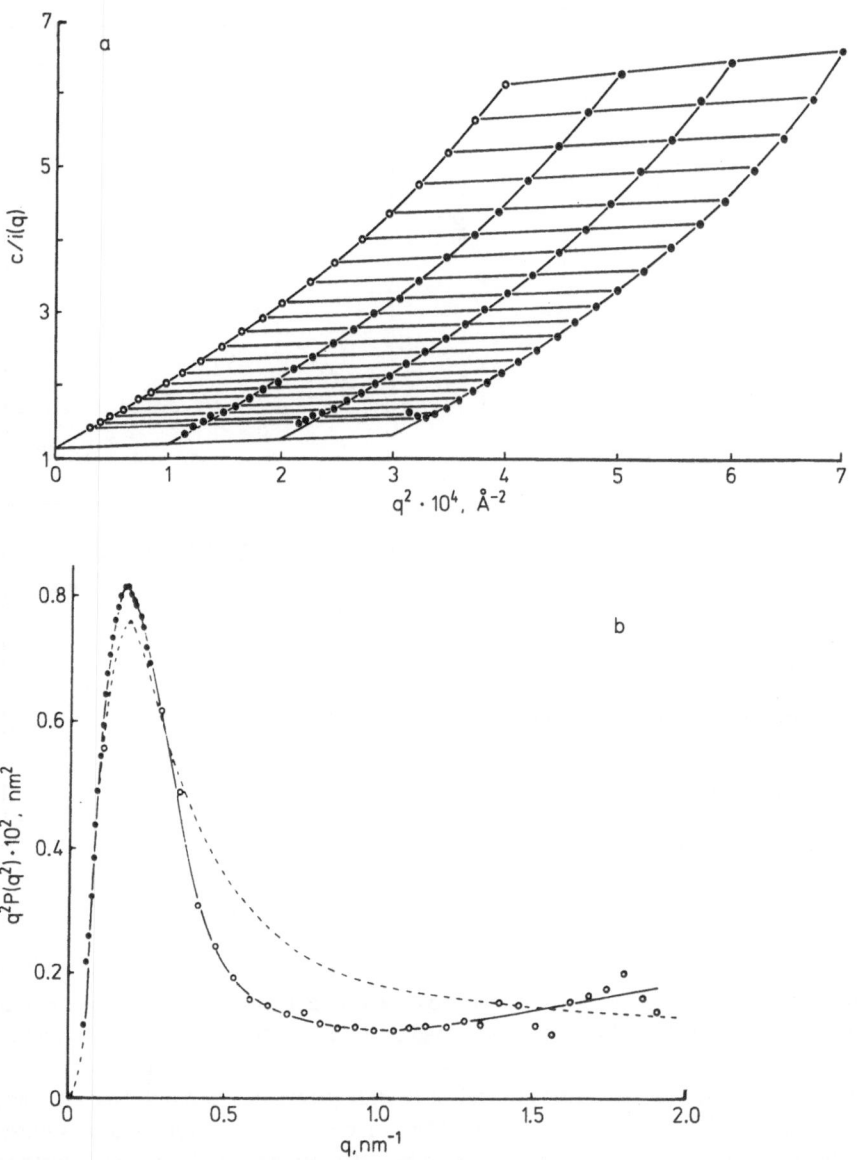

Fig. 30. Zimm-plot and Kratky-plot of glycogen from edible mussels. Neutron small-angle measurements in deuterated water. The *dotted line* describes the scattering behavior of the ABC model with $a_C = a_{AP}$, and a branching probability of $p = 0.25$[90, 174]

for special isomers and extending the result to the inherent polydispersity to derive $\langle S^2 \rangle_z$. In 1964 Gordon and Dobson[31] derived this z-average directly by applying a method by Kramers[11] to the cascade theory. The easiest way to derive $\langle S^2 \rangle_z$, however, results from the series expansion of the particle-scattering factor

$$\langle S^2 \rangle_z = -3\,[dP_z\,(q^2)/dq^2]_{q^2 = 0} \tag{D.22}$$

A list of various mean-square radii of gyration as function of the molecular weight is given in Table 3. In many cases this molecular weight dependence can be described by the approximation of a scaling law, i.e.

$$\langle S^2 \rangle_z \sim M_w^{2\nu} \tag{D.23}$$

The exponents are also given in Table 3. In most cases one obtains $2\nu = 1$ (no excluded volume effect is taken into account). Only in the two cases of the ABC polycondensates with the stringent constraint $\alpha_B + \alpha_C = \alpha_A$, and for the fractions of the random polycondensates, is the much lower exponent of $2\nu = 1/2$ obtained[90, 104] (see Fig. 31).

This lowered value of the exponent is considered by some authors to be a fundamental flaw of the FS theory, which neglects ring formation. The arguments by de Gennes[175] and Stauffer[44, 45] are as follows: near the critical point of branching, the molecular weight of the largest molecule tends to infinity. The segment density will increase with branching but can never exceed the value of closed packed spheres. Hence, near the critical point, $\langle S^2 \rangle$ of the largest molecule must increase at least with $M^{2/3}$. This argument is certainly correct for real chains where the finite volume of a segment does not allow two chain elements to be at the same position at the same time. In the present state of the FS theory the finite segment volume is mostly neglected, with the consequence that $\langle S^2 \rangle_z$ increases proportional to $M^{1/2}$ for monodisperse fractions, but excluded volume effects can be taken into account, and this was done by Kajiwara for small excluded volumes[34, 176].

Moreover, the overcrowding effect can be avoided in the cascade theory by introducing a second shell substitution effect. This was done by Gordon and Parker[177]. In the

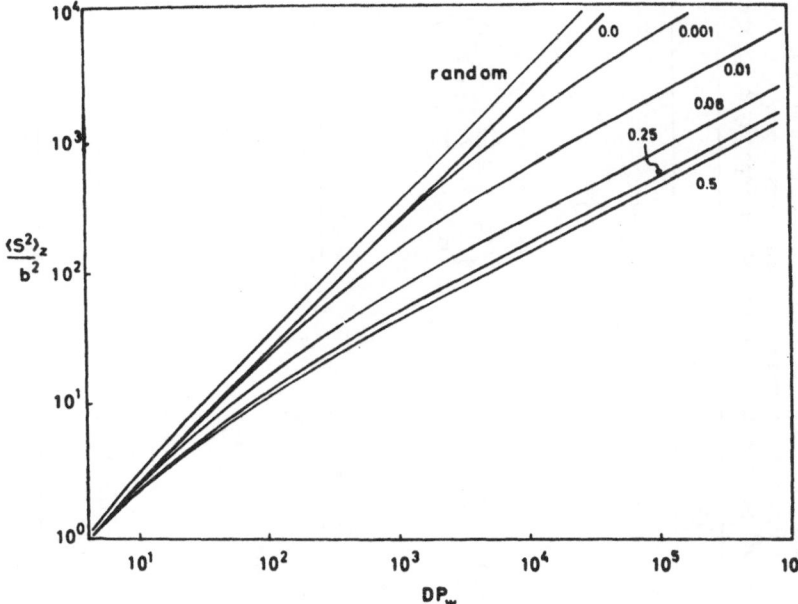

Fig. 31. Molecular weight dependence of $\langle S^2 \rangle_z$ for the ABC model. The numbers indicate the branching probability p in $\alpha_C = \alpha_A p$. The value $1/p$ gives the number of monomeric units between two branching points[104]

Table 3. Dependence of the mean-square radii of gyration $\langle S^2\rangle_z$ and the corresponding exponents ν_S and ν_D in the double-logarithmic plots of $\langle S^2\rangle_z$ and D_z vs. P_w for some selected models and the translational diffusion Constants D_z on the weight-average degree of polymerization P_w and the translational diffusion Constants D_z on the weight-average degree of polymerization P_w. ϱ factor and molecular polydispersity P_w/P_n for some selected models $\varrho = \langle 1/R\rangle_z \langle S^2\rangle_z^{1/2}$ ($A' = k_BT/6^{1/2}6^{1/2}\pi^{3/2}\eta_0 b$).

Model	$\langle S^2\rangle_z/b^2$	ν_S	D_z/A'	ν_D	ϱ	P_w/P_n
Linear chains						
Monodisperse	$\frac16 P_w$	1	$(8/3)/P_w^{1/2}$	$-1/2$	$8/3\,\pi^{1/2}$	1
Polydisperse (m = 1)	$\frac14 P_w$	1	$(2\pi/P_w)^{1/2}$	$-1/2$	$3^{1/2}$	2
Polydisperse (m coupled chains)	$\frac16\frac{m+2}{m+1}P_w^a$	1	$\left(\frac{\pi}{m+1}\right)^{1/2}2\sum_{k=1}^{m}\left(1+\frac{k-1}{m}\right)\frac{c(k)}{P_w^{1/2}}$	$-1/2$	$\frac{(m+2)^{1/2}}{m+1}2\sum_{k=1}\left(1+\frac{k-1}{m}\right)c(k)$	$1+(1/m)$
Star molecules						
Regular stars	$\frac16\frac{3f-2}{f^2}P_w^b$	1	$\frac83\frac{(2-f)+2^{1/2}(f-1)}{(fP_w)^{1/2}}$	$-1/2$	$\left(\frac{3f-2}{f\pi}\right)^{1/2}\frac83\frac{(2-f)+2^{1/2}(f-1)}{f}$	1
Polydisperse stars	$\frac{f}{(f+1)^2}P_w$	1	$\left(\frac{\pi}{f+1}\right)^{1/2}\frac{f+3}{2P_w^{1/2}}$	$-1/2$	$\left(\frac{6f}{f+1}\right)^{1/2}\frac{f+3}{2(f+1)}$	$1+(1/f)$
Polycondensates						
A_f type	$\frac{f-1}{2f}P_w^c$	1	$\left(\frac{\pi f}{(f-1)P_w}\right)^{1/2}$	$-1/2$	$3^{1/2}$	$P_w\left(1-\frac{f}{2(f-1)}\right)$
ABC type	$\frac{1+2B}{4(1+B)^2}P_w^d$	½ to 1	$\pi^{1/2}\frac{2+B}{[2(1+B)P_w]^{1/2}}$	$-1/4$ to $-1/2$	$\left(\frac34\frac{1+2B}{1+B}\right)^{1/2}\left(\frac{2+B}{1+B}\right)$	$2(1+B)$
Randomly cross-linked chains (polydisperse (m = 1) primary chains)	$\frac{P_w}{2(2+\varepsilon^*)}$	1	$\pi^{1/2}\left(\frac{2(1+\varepsilon^*)^{1/2}}{P_w}\right)$	$-1/2$	$3^{1/2}$	$2(P_w/P_wp)$

Model	$\langle S^2 \rangle_z/b^2$	ν_S	D_z/A'	ν_D	ϱ	P_w/P_n
Rigid and regular shapes						
Sphere	$\left(\dfrac{3}{5}\right)r^2$		$\dfrac{2}{2}$	$-1/3$	$(3/5)^{1/2} = 0.775$	1
Hollow sphere[a]	r^2		1	$-1/2$	1	1
Hollow sphere[b]	$\dfrac{3}{5}(r_0^5 - r_i^5)/(r_0^3 - r_i^3)$		$\dfrac{2}{3} < \epsilon < 1$	$-1/3$ to $-1/2$		1
Ellipsoid	$(a^2 + b^2 + c^2)/5$		$\dfrac{2}{3} < \nu_s < 2$	$-1/3$ to -1		1
Rod	$l^2/12$		2	$-1/3$ to -1		1
Disc	$(a^2 + b^2)/4$		1	-1		1
Cylinder	$\left\|a^2 + b^2 + 1\dfrac{2}{3}\right\|/4$		$1 < \nu_s < 2$	$-1/2$ to -1		1
Soft sphere	$b^2 \dfrac{3m^3}{2N^2}[X(1/3 + 10n - 6) + 6X^2(n-2) + 4n + (2(X+1)n - 3X)/m]^f$				$\begin{array}{c}1.4007\\ \text{to } 0.977\end{array}$	$\begin{array}{c}1\\1\end{array}$

[a] m = polydispersity parameter; $m^{-1} = P_w/P_n - 1$, where the subscripts denote weight- and number-average degrees of polymerization. [30]   [b] f = number of rays per star (functionality of the star center).   [c] f = functionality of the monomeric unit.   [d] $B = p(1-p)/(1-\alpha)$, where p is the branching probability and the link probability $[(1-\alpha)]^{-1} = P_n]$. B is very nearly the number of branching points per molecule based on the number-average degree of polymerization P_n.   [e] $\epsilon^* = \epsilon(n-1)x$, where ϵ is the probability that a pendant double bond has been involved in a cross-link. $n-1$ is the number of pendant vinyl double bonds per multifunctional vinyl monomer, and x is the molar fraction of multifunctional vinyl monomers in the branched molecule.   [f] $X = 2^n - 1$; n = number of shells, m = number of units per subchain, N = degree of polymerization

theory of substitution effect, the reactivity of the individual functional groups depends on how many of the functional groups had already reacted. In the nomenclature of genealogy, the fertility of a man depends on how many children the brothers of that man produced. The influence of this substitution effect on the critical exponent has not yet been investigated. It appears, therefore, unjustified to regard this as a flaw, in principle, of the FS theory.

We have to ask, however, how large an error is introduced when the excluded volume effect is neglected. Before considering this question, a recent result of cluster size distrition for antigen coated latex spheres which were cross-linked by antibodies, may be discussed. Schulthess et al.[178] measured this distribution as a function of the mean number of bonds per latex particle, $b = \alpha f$, where f is the number of antigens bound per latex particle and α the extent of reaction.

They found near $b_c = 1$

$$P_w \sim 1/(1 - b)^2 \sim \varepsilon^{-2} \tag{D.24}$$

where

$$\varepsilon = (b_c - b)/b_c = (1 - b/b_c)$$

which is exactly the result of the FS theory for the AB_2[111] model.

With the same model, one finds for the mean-square radius of gyration

$$\langle S^2 \rangle_z \sim 1/(1 - b) \sim P_w^{1/2} \tag{D.25}$$

which is the exponent criticized by de Gennes and Stauffer on theoretical grounds. However, here Eq. (D.25) follows immediately from the experimental size distribution.

At this point we have to ask why this low exponent of $\nu = 1/4$ is experimentally observed in contrast to the inevitably theoretical conclusion. Two facts have to be taken into consideration:

i)   First, the clusters of the largest molecular weight are present in the system with a very low weight fraction[7, 110]; most molecules are undercritically branched so that an $M^{1/2}$ dependence of $\langle S^2 \rangle_x$ of special clusters can be realized without difficulties since the density for these clusters is far below the critical close sphere packing.

ii) Second, the excluded volume effect in highly branched samples will be less than in a linear chain of the same degree of polymerization since not all segments can come into contact. Simple sterical constraints make this impossible. These constraints will already dominate after a few generations of branching, such that the excluded volume interaction has influence only within a certain correlation volume; beyond this radius, the thermodynamic interaction quickly loses its influence. These constraints are very probably the reasons why exponents of the unperturbed chains $2\nu = 1$ are often found in experiments[117, 137, 138, 179] but with dimensions which show the molecules as being swollen. The picture resembles more the shielding model by Edwards[180] than the blob-model by de Gennes and Daoud[103, 181, 182]; the latter appears in its deterministic character not to be well adapted to the statistical requirements.

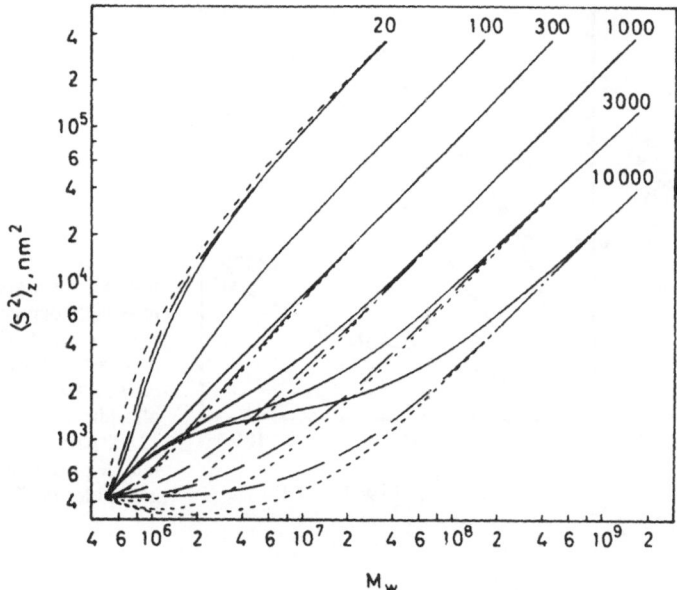

Fig. 32. Theoretically predicted molecular-weight dependence of the mean-square radius of gyration for star-molecules with rays grafted onto a large nucleus. *Full line:* nucleus is a "hard sphere"; *chain curve:* ABC nucleus; *dotted line:* A_3 nucleus[114]

Finally, the behavior of star-molecules with rays grafted onto an extended branched molecule may be discussed. Figure 32 exhibits the theoretically predicted molecular weight dependence of $\langle S^2 \rangle_z$ as a function of the number of grafted chains and the ray-length[114]. If only a few rays are attached to the nucleus, $\langle S^2 \rangle_z$ increases rapidly in the beginning and approaches a linear growth with M_w for large ray lengths. These asymptotic straight lines run parallel to that of an isolated linear chain but shifted towards lower values of $\langle S^2 \rangle$. This shrinking, usually described by the shrinking factor g[13], is a result of branching. If on the contrary a large number of rays are grafted onto the nucleus, the opposite behavior is observed, i.e. $\langle S^2 \rangle_z$ increases very slowly and may even decrease a little for short ray lengths. Again, in the limit of very long rays, a linear growth with M_w is predicted.

The predicted properties have indeed been observed experimentally with linear amylose chains which were synthesized by enzymes (potato- and muscle-phosphorylase) onto the nuclei of glycogen and partially degraded amylopectin[142–146]. The result is shown in Fig. 33 for runs with 15, 400 and 12000 rays, respectively. The feature is understood qualitatively as follows: if only a few chains are attached to the nucleus, the mass, compared to that of the nucleus, will increase only slightly while the dimensions grow rapidly with the ray length. The opposite is true for a very large number of rays; here the mass increases strongly while the dimensions remain almost unchanged. The slight decrease in $\langle S^2 \rangle_z$ becomes understandable if we take into consideration that rays can also emerge from free chain-ends which are positioned more deeply within the nucleus. Since

$$\langle S^2 \rangle = M^{-1} \sum_{j=1}^{x} \langle r_j^2 \rangle m_j \qquad\qquad (D.26)$$

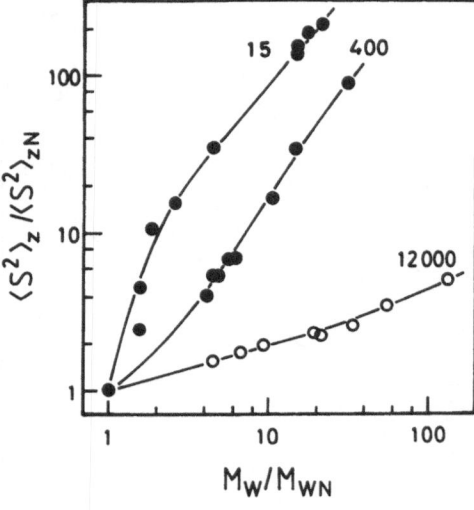

Fig. 33. Experimentally observed molecular-weight dependence of $\langle S^2 \rangle_z$ for amylose chains grafted onto glycogen (*filled circles*) and partially debranched amylopectin (*open circles*). Grafting was achieved by potato phosphorylase (15 rays for glycogen and 12 000 per amylopectin molecule), and by muscle phosphorylase (400 rays per glycogen molecule)[142, 143]

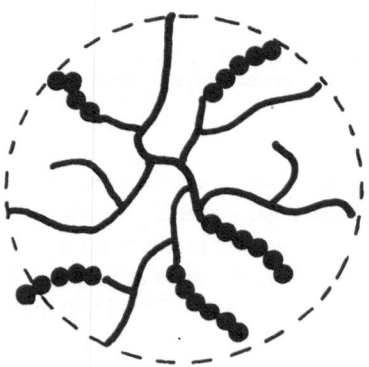

Fig. 34. Sketch of a model where the molecule increases in weight during a reaction but not in size resulting in a slight decrease of $\langle S^2 \rangle$. The filled circles indicate the grafted monomeric units

the mass will increase quickly while the equivalent sphere will grow only slightly in the beginning. Hence, the segment density is increased and $\langle S^2 \rangle$ must decrease. The situation is sketched in Fig. 34.

II. Dynamic Scattering Behavior

1. The First Cumulant

In Chap. B II.4 we have shown that the angular dependence of the first cumulant of the electric field correlation function can be obtained by integration over the particle-scattering factor. This rule remains valid also for copolymers but is restricted to Gaussian behavior of the subchains. Although the whole q-region can be covered by this integration, which in most cases has to be carried out numerically, it is useful to discuss the properties of the cumulant in three regions i.e. (i) $u \ll 1$, (ii) $u \simeq 1$, and (iii) $u \gg 1$, where $u = (\langle S^2 \rangle_z)^{1/2} q$.

(i) $u \ll 1$

This case was already considered in Chap. B II.4, Eq. (B.39), where the relationship (B.39) was derived

$$\lim_{q^2 \to 0} \Gamma/q^2 = D_z \tag{B.39}$$

with D_z being the z-average of the diffusion coefficient which is defined as

$$D_z = \frac{\Sigma w_j M_j D_j}{M_w} \tag{B.39'}$$

(ii) $u \gg 1$

This limit was already treated by de Gennes in 1967[183] on the basis of the hydrodynamic pre-average approximation. The correct asymptote was found by Stockmayer[82]. The basis of calculation is in both cases essentially the same: At large q-values only short sections are seen in a scattering experiment. (For the larger sections the pair-scattering function $\langle \exp(i q \cdot (r_j - r_k)) \rangle$ has decayed to very low values and does not contribute to the scattering intensity). Therefore, the first cumulant Γ or Γ/q^2 must become independent of the chain length and independent of molecular polydispersity. This criterion is sufficient for the calculation of the ν_g exponent as well as for the amplitude a in the asymptotic behavior

$$\Gamma/q^2 \to a q^{\nu_g} \tag{D.27}$$

In the following, a derivation of the exponent from scaling arguments is given[184] since this type of argumentation has become very popular in polymer physics, and since we wish to point out the limitation of this method.

First, the reduced cumulant Γ/q^2 must have dimensions of a diffusion coefficient and may therefore also be called an apparent diffusion coefficient D_{app}. We may write

$$\Gamma/q^2 = D_z f(\langle S^2 \rangle_z^{1/2} q) \tag{D.28}$$

where $f(u)$ is a function of a dimensionless quantity u. This parameter u must contain the variable q since it describes the angular dependence of Γ/q^2; q has the dimension of a reciprocal length, thus any characteristic length of the molecule could be chosen to form a dimensionless parameter. The choice of the radius of gyration is only a sensible one.

Second, an equivalent hydrodynamic radius can be defined from the diffusion coefficient via a Stokes-Einstein relationship

$$D_z = (kT/6\pi\eta_0) \langle 1/R_h \rangle_z \tag{D.29}$$

Thus

$$\Gamma/q^2 = (kT/6\pi\eta_0) \langle R_h^{-1} \rangle_z f(\langle S^2 \rangle_z^{1/2} q) \tag{D.28'}$$

So far no approximations have been made. In scaling theory the assumption is now made that $f(u)$ may be a homogeneous function, i.e.

$$f(u) \sim u^{\nu_g} \tag{D.30}$$

where nothing can be said about the amplitude. Inserting this relationship into Eq. (D.28'), one obtains

$$\Gamma/q^2 = (kT/6\pi\eta_0) \langle R_h^{-1} \rangle_z [\langle S^2 \rangle_z^{1/2} q]^{\nu_g} \tag{D.31}$$

Third, R_h and $\langle S^2 \rangle_z^{1/2}$ are functions of N of the same power. Therefore, the condition that Γ/q^2 must become independent of N, leads immediately to $\nu_g = 1$, and hence

$$\Gamma/q^2 \rightarrow (1/16)(kT/\eta_0) q \tag{D.32}$$

The amplitude factor of 1/16 was derived by Stockmayer[82] by solving Eq. (B.36) and (B.45) in the limit of large q, analytically. As already mentioned, the amplitude cannot be derived from scaling arguments.

(iii) $u \simeq 1$

The scaling argumentation has been outlined in some detail for the following reason: the asymptotic behavior is fairly easily, and also correctly, derived by this technique. Often, however, the result is extended into a region of smaller q-values, down to $u \simeq 1$[184], and now the scaling argument comes to a conclusion which deviates strongly from the result of the analytic solution of Eq. (B.45).

De Gennes, for instance, states that there will be a cross-over at about $u = 1$ from a q-independence at smaller u-values to a linear q-dependence at large u. This led experimentalists to draw two lines which form a kink at $u \cong 1$[184-187]. This conjecture can be shown to be wrong. Starting with Eq. (B.45), or with Eq. (B.49), the *Dawson integral*, or the exponential, can be expanded up to powers of q^2, and this truncated series can in most cases be summed analytically as required in Eq. (B.36). For linear chains (flexible monodisperse and polydisperse chains and rigid rods) and for branched structures of various kinds, the following general result is obtained

$$\Gamma/q^2 = D_z(1 + C \langle S^2 \rangle_z q^2 - ...) \tag{D.33}$$

where C is a characteristic constant that depends on the molecular architecture and polydispersity[102]. Eq. (D.33) shows that there is no direct change from a q-independence to a linear q-dependence but there is a region of at least a q^2-dependence in between. Figures 35, 36, 37 and 38 show some theoretical curves for $\Gamma/q^2 D$ as a function of q^2 and of q. Most striking is the result for the "soft sphere" model which exhibits a strong upturn

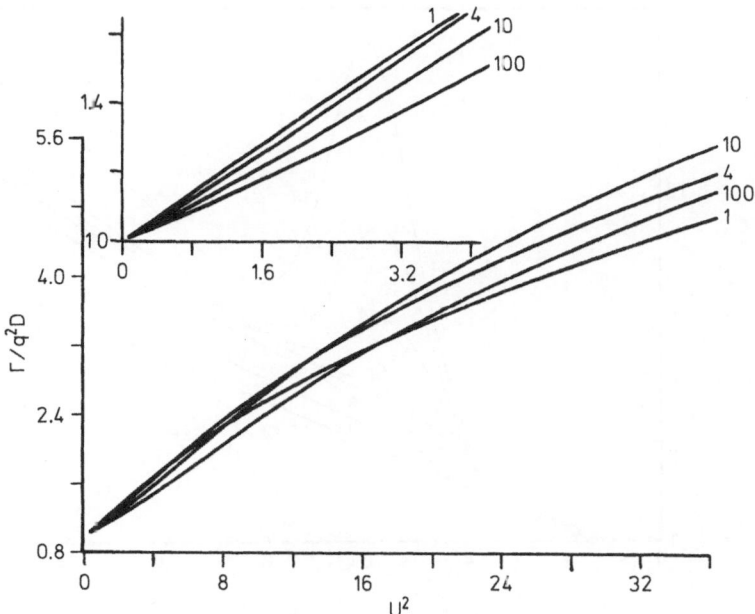

Fig. 35. Dependence of the reduced and normalized first cumulant Γ/D_q^2 on $u^2 = \langle S^2 \rangle q^2$ for regular stars. Insert: behavior at small u^2 [102)]

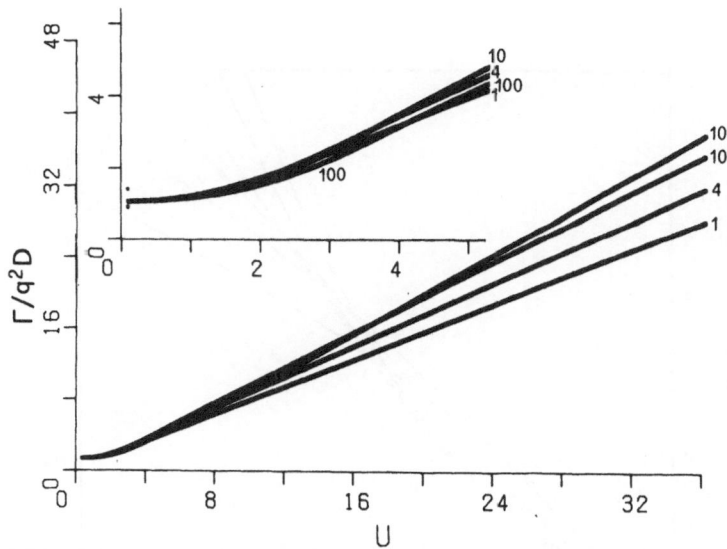

Fig. 36. Dependence of the reduced and normalized first cumulant $\Gamma/Dq^2 = D_{app}/D$ on $u = \langle S^2 \rangle^{1/2} q$ for regular stars[102)]

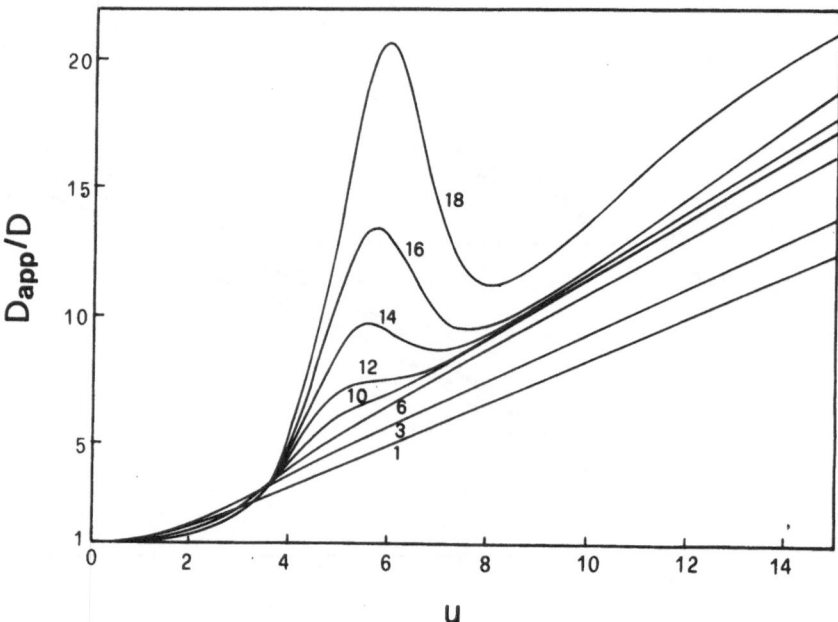

Fig. 37. Dependence of the reduced and normalized first cumulant $\Gamma/Dq^2 = D_{app}/D$ on u for the "soft sphere" model of different numbers of branching shells[93)]

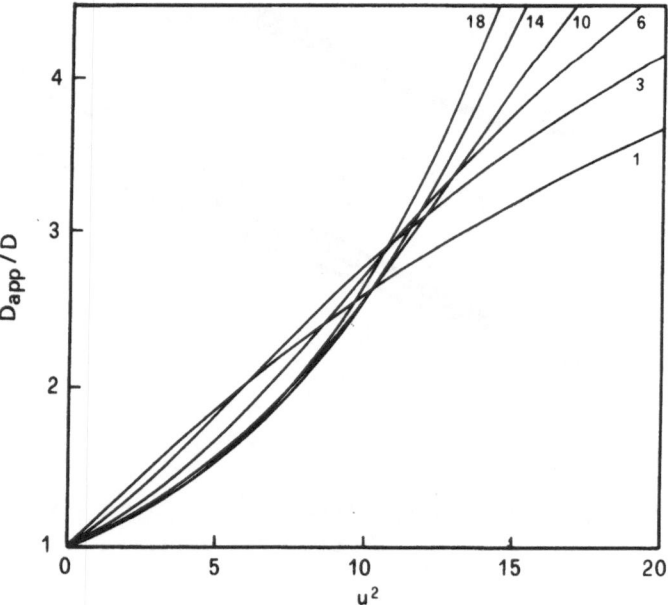

Fig. 38. Theoretically predicted u^2-dependence of the normalized first cumulant D_{app}/D for the "soft sphere" model[93)]

that becomes more and more pronounced with increasing number of branching shells[93]. Eventually a maximum builds up before the curve approaches the asymptotic linear q-dependence. This upturn has in fact been observed with microgels from PVAc[188, 189] as is shown in Fig. 39. The solid lines are theoretically predicted. A closer inspection of the curves for the regular star-molecules already reveals a weak sigmoidal shape for these structures[102].

One may ask the reason for the strong upturn in the regularly branched "soft sphere" model and the sigmoidal shape of the curves for the regular stars. It turns out that this behavior does not show up when the hydrodynamic pre-average approximation is applied[93]. As an example, the deviation $\Delta \Gamma/\Gamma$ from the correct cumulant is given for the regular stars in Fig. 35[82]. With increasing ray number, the deviations rise to 40% in an intermediate u-region and falls to 16% for large q, which is the deviation for the linear chain (see Fig. 40). The maximum deviation occurs at a point where the space correlation function $4\pi\gamma(r)rdr^2$ has its maximum density.

In summary, the analytic calculations lead to three relevant conclusions for branched structures:
(1) The application of the hydrodynamic pre-average approximation yields wrong results in quantitative respects, and sometimes also qualitatively
(2) The uncritical application of scaling arguments will result in a qualitatively incorrect picture
(3) The region around $\langle S^2 \rangle_z^{1/2} q \cong 1$ is extremely sensitive to structure and, thus, cannot be described by some laws of universality similar to those of the scaling or renormalization group theory. Moreover, this region contains the most valuable information on any particular system.

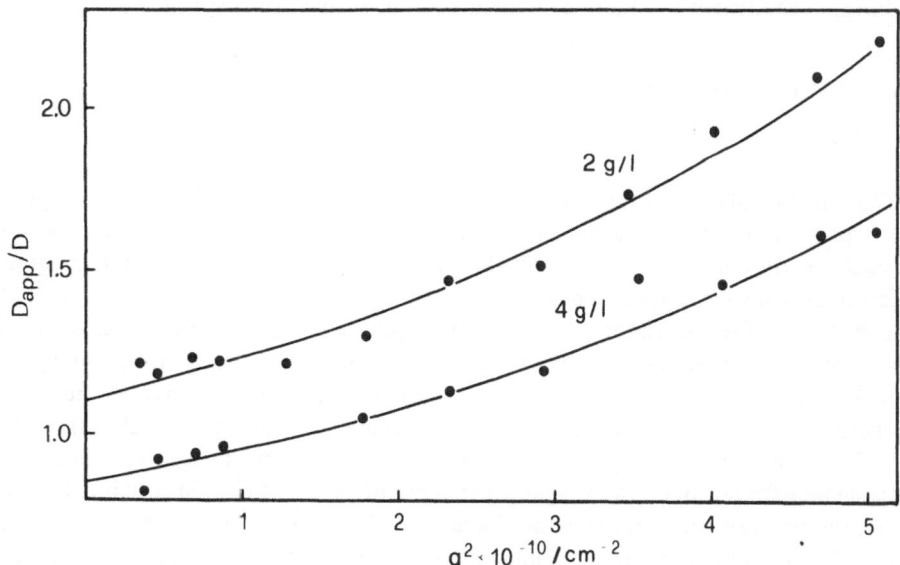

Fig. 39. Measurement of the apparent diffusion coefficient $D_{app} = \Gamma/q^2$ for two concentrations of a PVAc microgel in methanol[188, 189]. The full lines are theoretical curves for a "soft sphere" model with 7 branching shells[93]

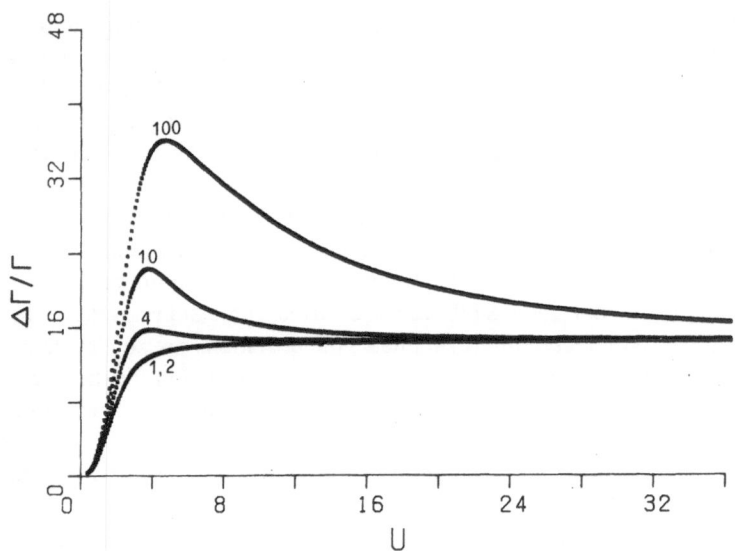

Fig. 40. Relative deviation of the first cumulant in the hydrodynamic pre-average approximation from the correct first cumulant for regular star-molecules of various ray numbers[82]

2. The Translational Diffusion Coefficient

While the mean-square radius of gyration $\langle S^2 \rangle_z$ can be obtained from the procedure of differentiating the particle-scattering factor $P_z(q^2)$ with respect of q^2, the translational diffusion coefficient is obtained by integration of $P_z(q^2)$ over the whole q^2 region[94]

$$D_z = (kT/3\pi^2\eta_0) \int_0^\infty P_z(q^2) \, dq \qquad (B.56)$$

This equation also holds for complicated copolymers where the particle-scattering factor is expressed in terms of matrices and vectors, but is limited to Gaussian behavior of the subchains. In a few favourable cases (linear chains and star-molecules), the integration can analytically be performed resulting in previously well established relationships[94]; but in most cases the integration has to be carried out numerically. Table 3 gives a list of D_z for various structures as function of the molecular weight and the corresponding exponents $-\nu_D$. The table also contains the hydrodynamic radius defined by Eq. (D.29) where, however, the free draining term is neglected (first term in Eq. (B.44)).

In common systems, D_z decreases with M_w and the exponent $-\nu_D$ is very close to the exponent ν for the molecular weight dependence of $\langle S^2 \rangle_z = M_w^{2\nu}$ if M_w is large. For small M_w, the free draining term must not be neglected and $-\nu_D$ often becomes smaller than ν[190-192]. In addition to this draining effect, Weill and des Cloiseaux[193] showed by a simple model calculation that $-\nu_D$ approaches its asymptotic value more slowly than does ν, if Gaussian statistics is perturbed by excluded volume effects. This results is a consequence of the fact that in

$$\langle 1/R_h \rangle = (2x)^{-1} \sum_{j \neq k}^{x} \sum^{x} \langle 1/r_{jk} \rangle$$

the shorter chain sections are more strongly weighted than in

$$\langle S^2 \rangle = (2x)^{-1} \sum_{j}^{x} \sum_{k}^{x} \langle r_{jk} \rangle^2$$

where the larger subchains are weighted. The excluded volume effect, however, increases more steeply viz. $\propto n^{1/2}$, where n is the number of segments which form a subchain path, and the excluded volume effect is thus smaller for short subchains than for large ones.

Polyvinyl acetate polymers synthesized by emulsion polymerization[168], show strikingly different behavior[138, 189]. Figure 41 shows D_z as a function of M_w. For the low molecular weights, D_z decreases as expected but from $M_w = 14 \cdot 10^6$ onwards D_z remains constant. The measured samples were products obtained at different extents of monomer conversion. The soap surrounding the latex particles was removed and the samples were dissolved in methanol and measured. The behavior of D_z in this system is understood by recalling that

(i)   the polymerization of vinyl acetate results in highly branched polymers as a result of the high chain transfer reaction of the radical, due to the acetate groups, and

(ii) that the branching process occurs within the fairly small volume of a latex sphere of about 80 nm in radius.

At low monomer conversion, the polymerization leads to fairly small molecules, and the branching process can proceed largely unimpeded by the finite volume. Thus, the common behavior of randomly branched polymers is observed. At larger conversion of monomer, the polymer has grown in size, such that the largest species have already reached dimensions of the latex sphere. On further branching, the molecular dimensions

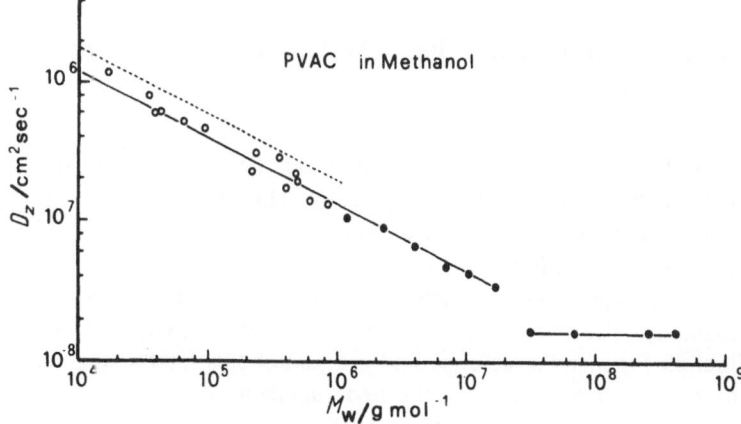

Fig. 41. Molecular weight dependence of D_z of linear PVAc (O), and of branched PVAc synthesized in latex particles (●)[188, 189]

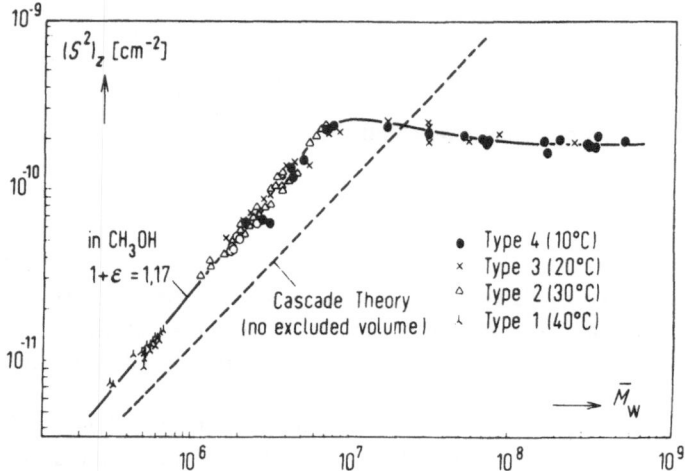

Fig. 42. Molecular weight dependence of $\langle S^2 \rangle_z$ for PVAc polymerized in latex particles[168, 189]

of the largest molecules cannot grow further in size. Further attachment of chains to these molecules, which leads to an increase in their segment density, can occur only if simultaneously other, smaller molecules can grow up to the size of the latex sphere and thus enlarge the gel fraction. In other words, gelation takes place in the latex particle where the spherical shape becomes preserved and the amount of molecules, smaller in size than the latex particle, will decrease. Since D_z is related to the hydrodynamic radius via the Stokes-Einstein relationship (see Eq. (D.29)), and since the radius of the microgel particle remains almost constant after gelation, the diffusion coefficient of these microgels must also stay constant. Figure 42 shows that similar behavior was found for the mean-square radius of gyration[168, 189] which was measured by static light scattering.

III. Information from the Combination of Static and Dynamic Light Scattering[102]

In the last two chapters some major properties of the static and dynamic scattering functions have been discussed separately. This chapter deals with the combination of both techniques and with the question of whether such a combination can produce additional information.

In fact, only a fairly restricted field can be covered if the two techniques are applied separately. The molecular weight dependences of $\langle S^2 \rangle_z$ and D_z are not particularly different for many structures, and the same holds for the angular dependence of the particle-scattering factor and the first cumulant of the time-correlation function. The situation is especially disappointing for biological products, where often different molecular weights cannot be obtained without destroying the structure. A similar situation exists with systems that change their properties with time.

1. The Geometric and Hydrodynamic Shrinking Factors

In some cases, one step forward can be made by comparing $\langle S^2 \rangle_z$ and D_z with the corresponding quantities of a linear chain of the same composition and the same molecular weight. In their pioneering papers, Stockmayer, Zimm and Fixman defined two shrinking factors g_z[13] and h_z[91] for branched molecules based on the ratio of radii $\langle S^2 \rangle_{zb}/\langle S^2 \rangle_{z\ell}$ and $R_{gb}/R_{h\ell}$ of a branched and a linear molecule of the same z-average molecular weight M_z. These two g- and h-factors have become quite popular, but a direct application is limited by the fact that M_z is not a directly measurably quantity, and moreover that the linear and branched molecules must have the same molecular weight distribution. In common polymerization techniques, these conditions are never met; randomly branches polymers exhibit pathological broad molecular weight distributions with polydispersities of M_w/M_n of 20 to 100 or even more. Such broad molecular weight distributions can never be achieved with linear chains. Therefore, some mathematical manipulations have to be made before the formulae for g_z and h_z can be used. These techniques are discussed extensively by Hoffmann et al.[194].

In 1980, the g- and h-factors were defined in a slightly different[102] manner making the direct application more feasible

$$g = \langle S^2 \rangle_{zb}/\langle S^2 \rangle_{z\ell} \quad | \quad \text{at the same } M_w \tag{D.34}$$

$$h = R_{hb}/R_{h\ell} \quad | \qquad \text{at the same } M_w \tag{D.35}$$

No restriction is made to the same molecular weight distribution. Instead of this, the natural distributions for $f > 2$ and $f = 2$ are taken. For star-molecules, $f = 2$ corresponds to the monodisperse linear chain or to a linear chain that obeys the most probable distribution, and in the case of random polycondensates, $f > 2$ corresponds to the branched non-fractionated sample, and $f = 2$ to the linear polycondensate. The g and h-factors so defined no longer have the appearance of shrinking factors in all cases, as may be recognized from Figs. 43 and 44. For star-molecules, both factors decrease as

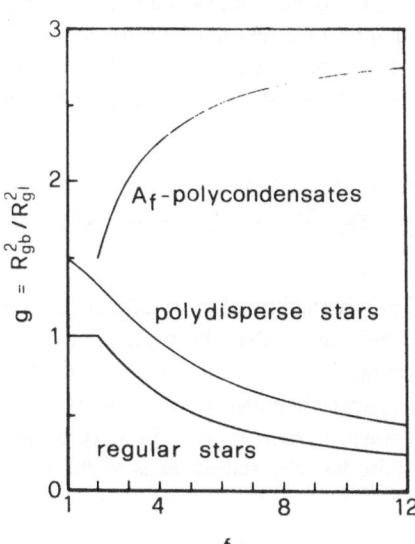

Fig. 43. Dependence of the g-factor, defined by Eq. (D.34), on the functionality f for random polycondensates, and on the ray number f for star-molecules[102]

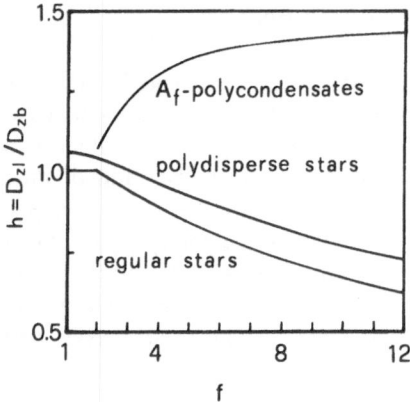

Fig. 44. The same as in Fig. 43 but for the hydrodynamic factor h defined by Eq. (D.35)[102]

expected, but for the random polycondensates a continuous increase with the functionality number of the monomers is observed. The decrease of g and h for regularly branched structures has a simple explanation: if several chains are attached to one branching point, then the mass increases considerably while the radius of gyration varies only slightly; or in other words, to keep the same molecular weight, the attached chains must be correspondingly shorter resulting in a lower $\langle S^2 \rangle$ of the branched molecule. However, if the branched molecules have a much broader molecular weight distribution than the linear standard, then the z-average of the mean-square radius of gyration increases at the same M_w because of the polydispersity, and it decreases as a result of branching. In the randomly branched polymers, the increase in polydispersity overcompensates the shrinking due to branching[34]. The definitions fo Eq. (D.34) and (D.35) have thus the advantage that randomly branched chains can now be distinguished from other more regularly branched structures, which is not the case if the original g_z- and h_z-factors are used. These always decrease with branching in a similar manner.

It may be emphasized here that the h-factor has become the more reliable parameter in spite of a less marked effect since with the recent photon-correlation spectroscopy, translational diffusion coefficients can be measured with a high accuracy and over the whole molecular weight range, from the monomer up to the largest molecular weights.

2. *The Parameter* $\varrho = \langle S^2 \rangle_z^{1/2} \langle R_h^{-1} \rangle_z$ [188, 102]

The characterization of branched or cyclic structures by the g- and h-factors has the disadvantage that the properties of the analogous linear chains must be known. Such characterization requires a great deal of work in preparative chemistry as well as accurate physical chemical measurements, and in some cases the linear analogs are not even known. In such cases, a direct combination of the mean-square radius of gyration and the hydrodynamic radius leads to the very useful dimensionless parameter[188].

$$\varrho = \langle S^2 \rangle_z^{1/2} \langle R_h^{-1} \rangle_z \tag{D.36}$$

The values for monodisperse linear [73, 195–197] and cyclic chains[188, 198] have long been known. Recently, calculations have been extended to polydisperse linear chains[200], which obey the *Schulz-Zimm type*[123, 156, 199] of molecular weight distribution[201], and to a number of branched structures[102]. The result is collected in Table 4, and some of the ϱ-functions are shown in Figs. 45 and 46 with the dependence on the functionality number of the monomeric units or the attached number of rays, respectively. One realizes from Figs. 46 a and b that polydispersity causes an increase of ϱ but branching a decrease. In randomly branched chains both effects balance, and no dependence on the functionality number becomes detectable.

The effect of excluded volume on ϱ for linear chains has been calculated, first more qualitatively by Weill and des Cloiseaux[193] on the basis of scaling arguments, then by Akcasu and Benmouna[202] quantitatively on the basis of the blob-model. The result is as follows

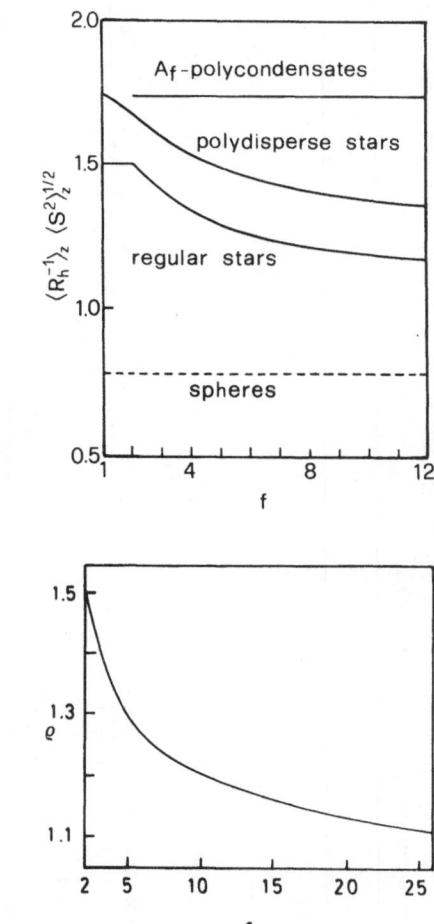

Fig. 45. The ratio $\langle R_h^{-1} \rangle_z \langle S^2 \rangle_z^{1/2} \equiv \varrho$ for various branched structures: f denotes the functionality of the monomers or the ray number per star-molecule, respectively[102]

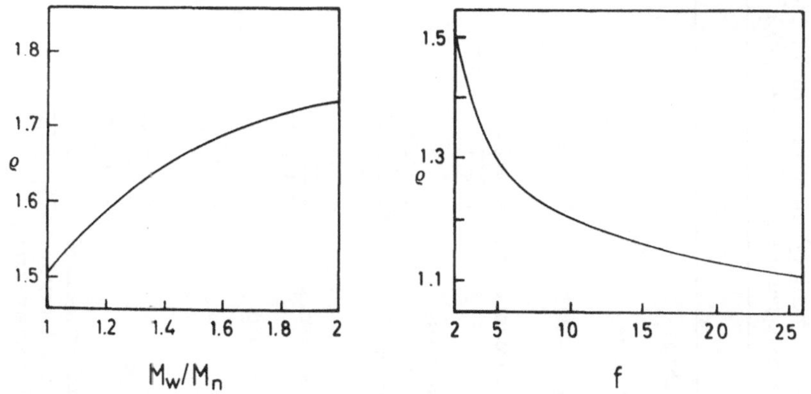

Fig. 46 a, b. Dependence of ϱ: (**a**) on the chain polydispersity for linear chains, (**b**) on the ray number for regular stars[35, 102]

Table 4. Geometric (g) and hydrodynamic (h) factors and coefficient C in the initial slope of the first cummulant for some selected models. C is defined through the equation $\Gamma/q^2 = D_z(1 + C\langle S^2\rangle_z q^2 - \ldots)$.

Model	g	h	C
Linear chains			
Monodisperse	1	1	$13/75 = 0.1733$
Polydisperse (m = 1)	3/2	$\frac{4}{3}\left(\frac{2}{\pi}\right)^{1/2}$	1/5
Polydisperse (m coupled chains)	$\frac{m+2}{m+1}$	$\frac{4}{3}\left(\frac{m+1}{\pi}\right)^{1/2}\bigg/\left(\sum_{k=1}^{m}\left(1-\frac{k-1}{m}\right)c(k)\right)^{a}$	$\frac{(13m+32)}{75(m+2)}$
Star molecules			
Regular stars	$\frac{3f-2}{f^2}$	$\frac{f^{1/2}}{(2-f)+2^{1/2}(f-1)}$	$\frac{1}{3}-\frac{4f}{25(3f-2)}\dfrac{3f+2^{1/2}f-2}{f+2^{1/2}}$
Polydisperse stars	$\frac{6f}{(f+1)^2}$	$\frac{16}{3(f+3)}\left(\frac{f+1}{\pi}\right)^{1/2}$	$\frac{(2f^2+11f-1)}{15f(f+3)}$
Polycondensates			
A_f type	$\frac{3(f-1)}{f}$	$\frac{8}{3}\left(\frac{f-1}{\pi f}\right)^{1/2}$	1/5
ABC type	$\frac{3}{2}\dfrac{1+2B}{(1+B)^2}$	$\frac{8}{3}\dfrac{[2(1+B)]^{1/2}}{\pi^{1/2}(2+B)}$	$\frac{1}{6}\dfrac{1+B}{1+2B}\left(1+\frac{1}{5}\dfrac{2+3B}{2+B}+\dfrac{B}{(1+2B)(2+B)}\right)$
Randomly cross-linked chains (polydisperse (m = 1) primary chains)	$\frac{3}{2(1+\varepsilon^*/2)}$	$\frac{8}{3}\left(\dfrac{1}{2\pi(1+\varepsilon^*)}\right)^{1/2}$	

$$B = \frac{p(1-p)}{1+B}\frac{P_w}{2}$$

a For the definition of c(k) see Ref. 30.

$$\varrho(v) = 6/\{(1 - v)(2 - v)[3\pi(1 + v)(1 + 2v)]^{1/2}\} \tag{D.37}$$

Equation (D.37) was derived on the assumption that for short chain sections, up to a certain "cross-over" length n_c, no excluded volume effect can exist at all; beyond n_c a maximum excluded volume characterized by the exponent $v = 0.6$ is assumed to be effective. To include experimentally observed lower exponents resulting from lower excluded volumes, the authors were forced to take n_c as an adjustable parameter, i.e. n_c was assumed to increase in length as the excluded volume is lowered; under θ-conditions n_c tends to infinity. Equation (D.37) must therefore be considered as a semiempirical relationship only. More exact calculations have been carried out by Stockmayer and Tanaka[203] for small excluded volumes on the basis of a perturbation theory.

The first experimental verifications of Eq. (D.37) have been made for linear PS[192, 204] PMMA[190, 191] and PDMS[205] chains and for cyclic PDMS[205] by a combination of static and dynamic light-scattering, or static and dynamic neutron-scattering. For branched chains, the ϱ-ratio was measured for epoxide resins[206, 207], sub-critically cross-linked polystyrene[208], and polyvinyl acetate (PVAc) emulsion polymerizates[188, 189].

The measurements from linear chains revealed two unexpected results. First, the ϱ-parameter was found to be 16–23% lower than predicted by theory both in θ-solvents and in good solvents[190–192, 204–209]. In other words, the hydrodynamic radius is larger than theoretically expected. Second, the ϱ-parameter decreases for lower molecular weights; the effect is more pronounced for the polymers in a θ-solvent than in a good solvent.

The experimentally observed lower value for ϱ is strong evidence for the *Oseen tensor approach* being an insufficient description of the hydrodynamic interaction. It has been known for a long time that Oseen's first order approximation for the hydrodynamic force between two spheres holds only for fairly long distances[210]. Several suggestions have been made in the past to improve this approximation[211–213], but all of them lead to an increase of ϱ instead of the observed decrease. Evidently, the hydrodynamic interaction is stronger than predicted by all these theories, so that the solvent becomes immobilized down to rather low segment densities, thereby producing a larger effective hydrodynamic radius. A closer examination at the hydrodynamic theories of flexible macromolecules shows that the over-all hydrodynamic interaction of the N beads in the coil is approximated by the sum over all interacting pairs of elements. The magnitude of this pair interaction is assumed to be independent of the relative position of a third or other units in the neighbourhood. Figure 47 demonstrates, what is meant, with a three bead model. Possibly, when the three-bead model has been solved exactly, an improvement will be achieved[214].

Similar deviations of the experimentally determined ϱ-value from theory have been observed with branched epoxides[206, 207]. The value of about 1.48 is close to that of monodisperse linear chains in a θ-solvent, but considerably lower than 1.73 predicted for randomly branched chains under θ-condition. Since the measurements were made in a

Fig. 47. Schematic representation of the hydrodynamic interaction (HI (1)) of two other beads 2 and 3 on a bead 1. The total HI may be written as HI (1) = h (1, 2) + h (1, 3) + [h (2, 3)] where h (1, 2) denotes the force of bead 2 on bead 1, etc. In the Oseen approach, the interaction between beads 2 and 3 is assumed to have no influence on bead 1

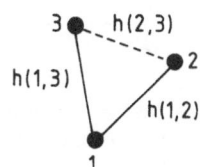

very good solvent, an even higher value for ϱ could be expected if we assume that the excluded volume has a similar influence on *branched* chains as on *linear* ones.

The measurements with PVAc molecules[188, 189] obtained from emulsion polymerization clearly disclose the great value of the parameter ϱ for structure determination. Figure 48 shows the result of ϱ-determination at various molecular weights. At low M_w, where essentially randomly branched chains are obtained, ϱ varies only slightly with molecular weight and has the typical value of randomly branched chains. Around $M_w = 14 \cdot 10^6$, a steep transition to a rather low value occurs, which then remains constant on further increase in M_w. Evidently, a change in structure occurs around $M_w = 14 \cdot 10^6$, and, in fact, at about that molecular weight, gel-formation inside of the latex particle was deduced from other measurements. The ϱ-value of the microgels lies well below the "hard sphere" value. The effect may be interpreted by the dangling chains of the microgels, which produce a much softer decay in the segment density, to larger radii than a "hard sphere" with its well defined surface. Again, a larger hydrodynamic radius or a low

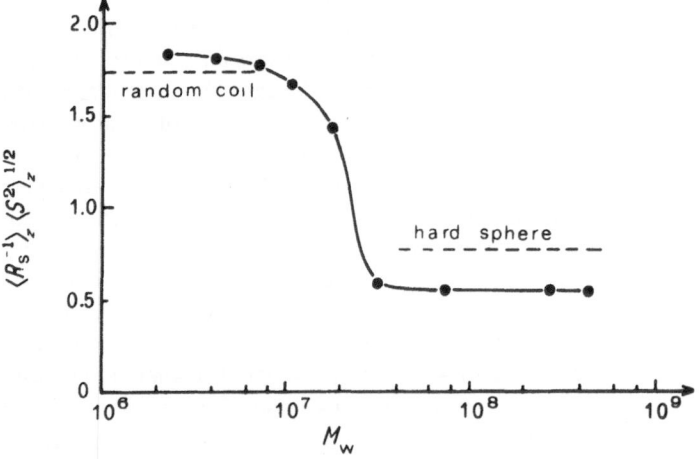

Fig. 48. The ϱ-parameter for PVAc obtained by emulsion polymerization. The transition indicates gel-formation in the latex particle. (Compare also Figs. 41 and 42[188, 169])

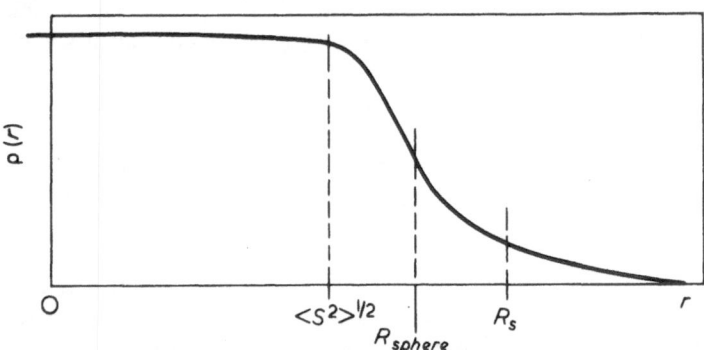

Fig. 49. A sketch of the segment density profile for a microgel. The various bars indicate the position of the radius of gyration, the expected equivalent sphere radius and the hydrodynamic radius, respectively[188]

ϱ-value can be expected if we assume that the solvent becomes immobilized down to a rather low segment density (see Fig. 49).

The observed decrease of ϱ at low molecular weights is very likely to be due to an influence of the free draining term, which has been neglected hitherto but which becomes noticeable for short chains. For chains in a good solvent one could also think of the *Weill-des Cloiseaux calculations* which show a less marked increase of the hydrodynamic radius at small M_w than for the geometric radius of gyration. These calculations would imply an *increase* of ϱ with falling M_w and, in fact, the experimentally observed decrease is weaker in a good solvent than in the θ-solvent. Evidently, the decrease in ϱ as a result of the free draining contribution becomes partly compensated by the excluded volume effect as predicted by Weill and des Cloiseaux[193]. We have to wait, however, for more experimental results before a clearer picture can be drawn.

The observations with the PVAc molecules now allow a tentative interpretation of the behavior of undercritically cross-linked polystyrene chains[208] measured by Kajiwara, Gordon and Charlesby[157]. The cross-linked materials were prepared by γ-ray irradiation of a powdered polymer in vacuum. Nearly monodisperse linear polystyrenes were used. Figure 50 shows the result of ϱ as function of γ/γ_c; γ_c is the critical γ-ray dose where gelation took place. One notices a slight increase in ϱ with branching but near the gel point a sharp decrease occurs. Evidently marked ring formation takes place at higher doses which give the cross-linked molecules the appearance more and more of a "soft sphere"[93] with surface better defined than that of a randomly cross-linked molecule.

3. The Coefficient C in Eq. (D.33)[35, 82, 102, 188, 189, 215]

As was already pointed out, the coefficient C in the q^2 dependence of the reduced first cumulant Γ/q^2 is a valuable measure of the structure. Theory predicts an increase of C with polydispersity but a decrease with branching. Thus, the behavior resembles that of ϱ but the dependence is different in quantitative respects[102]. The combination of both parameters C and ϱ therefore, should make possible, a better differentiation between various models. At present, only a few, partly contradictory, results are known for linear chains[184-187, 216, 217] and no data are available for branched molecules. The discussion of C as a structure-significant parameter must be kept short, not only for this reason but also because of experimental difficulties[209, 218]. These difficulties become evident when the experimental results from various laboratories are compared. With the exception of

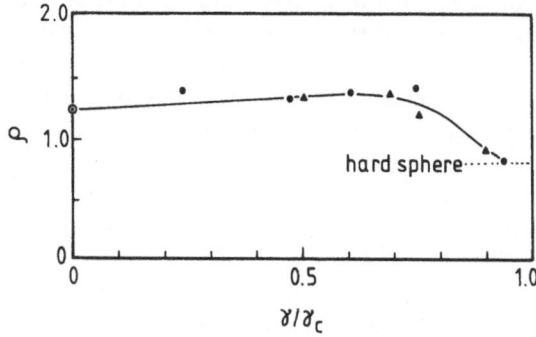

Fig. 50. Dependence of the ϱ-ratio for cross-linked polystyrene chains. γ_c denotes the critical γ-ray dose where gelation was observed, and γ/γ_c is the relative γ-dose. Various monodisperse linear chains have been used as starting materials. The measurements were made in cyclohexane under θ-conditions. The dotted line describes the theoretically predicted behavior[208]

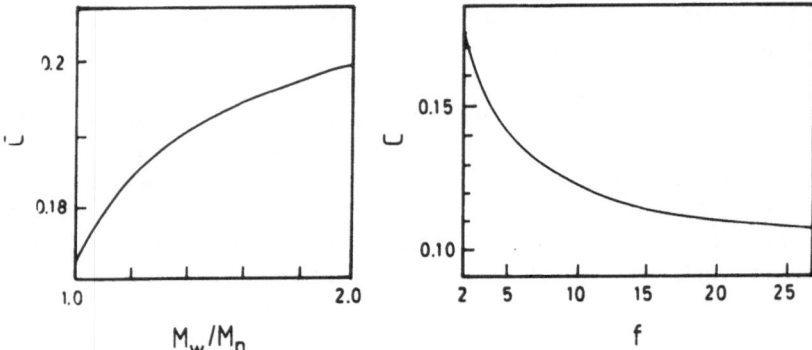

M_w/M_n f

Fig. 51. Increase of the coefficient C with molecular polydispersity for linear chains (*left*) and decrease of C with branching for regular star molecules (*right*)[102] predicted by theory

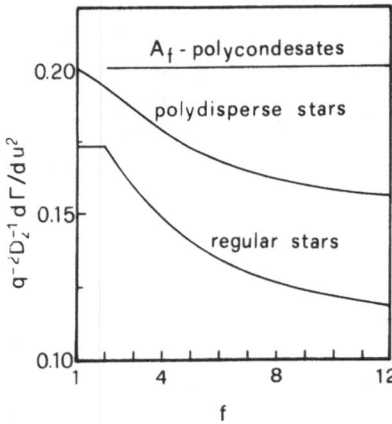

Fig. 52. The dependence of the coefficient C on branching for various branching models; f denotes the functionality of the monomer for the polycondensates, and of the star centre, respectively[102]

Caroline and his coworkers[26], all other laboratories found considerably lower values for C than theory predicts under the recommended optimum conditions. Figures 51 and 52 give a few examples for C according to theory[102].

In the treatment of a rigid dumbbell, where the whole time-correlation functions (TCF) can be solved exactly, Stockmayer and Burchard[219] disclosed the origin for the discrepancy between theory and experiments. They recognized that all measurements of the TCF can be carried out down only to a limiting minimum delay time. With common instruments, this lower limit lies at about 100 ns but the lowest time is often much higher under conditions such that the TCF should have decayed to e^{-2} at channel 80[220]. These experimental condition imply that only an apparent first cumulant is determined defined by

$$\Gamma(t_0) = \partial \ln D(q, t)/\partial t|_{t = t_0} \tag{D.38}$$

where t_0 is the delay time of the first channel of the autocorrelator; the correct first cumulant, however, requires the initial slope at $t = 0$.

$$\Gamma = \partial \ln S\,(q,\,t)/\partial t|_{t=0} \tag{D.39}$$

If the TCF is not a single exponential, the apparent first cumulant $\Gamma\,(t_0)$ will in general be smaller than Γ; only for very low q^2, does the TCF always become a single exponential. Consequently, at small q^2 the correct Γ is measured while with increasing q^2, where deviations from a single exponential become effective as a result of the internal modes which now contribute, $\Gamma\,(t_0) < \Gamma$. Hence, the coefficient $C\,(t_0)$ in the q^2 dependence of the apparent first cumulant $\Gamma\,(t_0)/q^2$ must be smaller than the theoretically predicted C value. The effect has been calculated by Stockmayer and Schmidt for rigid and Gaussian dumbbells[221], and for flexible linear chains on the basis of the Pecora model[218]. Figure 53 gives an example.

Fig. 54 shows some first systematic measurements by Bantle et al.[207, 209], who changed the delay time by factors of $1:2:4:8$; the suggested optimum condition for a recording of the full correlation time corresponds to a value of 4, i.e. the TCF has decayed here to e^{-2} of the original value at channel 80[220]. Evidently, the q^2 dependence can in these four cases be represented by

$$\Gamma\,(t_0)/q^2 = D_z\,(1 + C\,(t_0)\,q^2 - \ldots) \tag{D.40}$$

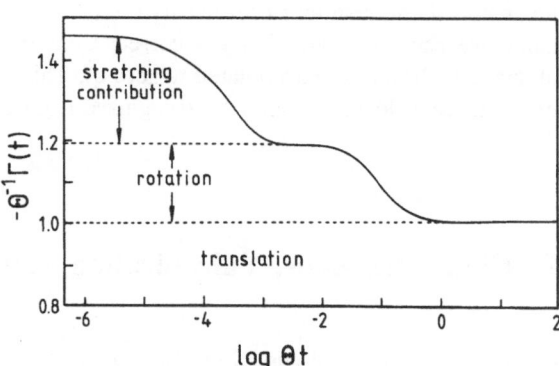

Fig. 53. Computed decay of the dynamic scattering function for a rather stiff dumbbell according to Ref.[219]. The rotatory diffusion coefficient is Θ, $qL = 2$, $a = L/50$, $\tau = \Theta/2500$. L = length of the dumbbell, a^2 = mean square amplitude of the bond stretching, $\tau = a^2/4\,D$ stretching relaxation time[221]

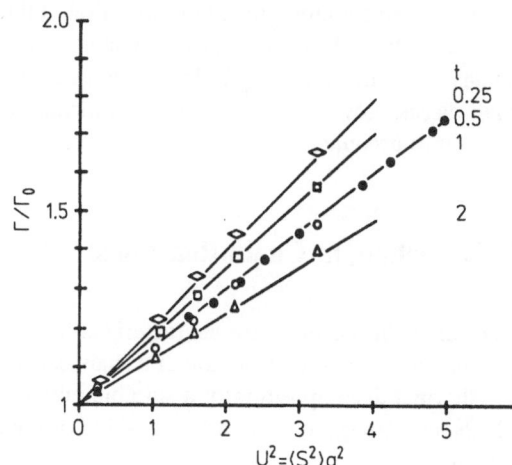

Fig. 54. The dependence of the first cumulant $\Gamma\,(t_0)/q^2\,D$ on the delay time t_0. $\Gamma_0 = q^2 D$[207, 209]

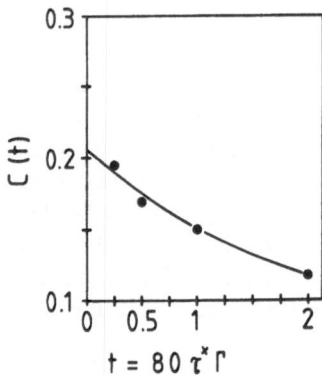

Fig. 55. Measurements of $C(t_0)$ at four different delay time t_0 for the first channel. The extrapolation results in a value which is close to the theoretically predicted one of $C = 0.183$ for the polydispersity of the polystyrene sample[207, 209]

Figure 55 demonstrates that, if the molecular polydispersity is taken into account, the apparently low value of $C(t_0)$ at the recommended t_0-time approaches very satisfactorily the correct value of C after extrapolation to $t_0 = 0$.

We thus come to the conclusion that the correct value of the coefficient C for structure determination can be gained only if the measurements were made at different delay times and if the resultant coefficients $C(t_0)$ are extrapolated to $t_0 = 0$. On the other hand, we can also state that the correct translational diffusion coefficient is always obtained if $\Gamma(t_0)/q^2$ is extrapolated to zero q^2. This is because the coefficient $C(t_0)$ loses its influence at low q^2, as may be recognized from Eq. (D.40), and from Fig. 54.

E. Chain Reactions, Vulcanization, Inhomogeneities

In Chap. C we have discussed in some detail the application of the cascade theory to polycondensates in their unperturbed state. In Chap. D some experimental results were already given for cross-linked or vulcanised linear chains. In this chapter we shall now outline in brief how cross-linking chain reactions or the vulcanization of preformed chains of an arbitrary length distribution can be treated by cascade theory. Second, we shall discuss how heterogeneities in branching or a rigidity of a certain domain can be taken into account.

I. Branching in Chain Reactions[106, 107, 116, 117]

Radical chain reactions are characterized by the coupled three stage reaction
(1) *initiation*, i.e. the formation of a chain carrier, mostly the creation of free radicals by thermal decomposition of a suitable initiator,
(2) *chain propagation*, i.e. a fast reaction of the activated chain carrier with monomers, and

(3) *termination* either by radical recombination or by chain transfer, where the chain growth is terminated by the transfer of a hydrogen atom or a chlorine atom to the radical by a suitable chemical reagent. After this chain transfer, the transfer reagent becomes itself a radical which is able to start a new polymer chain.

Branching, or *cross-linking* occurs if the polymer or the monomer bears functional groups which are capable of chain transfer. If the polymer group undergoes chain transfer, then a new chain propagates as a side chain from the polymer; if a group of the monomer performs the chain transfer, a linear chain with a terminal double bond is produced in a first stage, at high monomer conversion; i.e. if almost all of the monomeric double bonds are consumed, these terminal double bonds become polymerized into a growing chain, thus producing excessive branching.

To treat such chain reactions, we have to be aware that the polymerization proceeds always from a point of initiation to a point of termination. Correspondingly the process has to be described by a *directed graph*. Figure 56 shows as an example a rooted tree from a linear chain, where the arrows indicate the direction of the chain growth. In setting up the probability-generating function of a tree, we always have to ascend the tree from the root to higher generations.

Evidently, a monomeric unit chosen as the root of a tree has two *statistically unlike* functional groups, but of the *same reactivity*. Hence, the generation function for the zero-th generation is the product of the two functional group generating functions $F\downarrow$ and $F\uparrow$

$$F_0(s) = (1 - \alpha + \alpha s\downarrow)(1 - \alpha + \alpha s\uparrow) \tag{E.1}$$

and for the higher generations, we have to take care that the ascending to higher generations proceeds exclusively either *in* the direction of the polymerization direction or *against* this direction.

$$\begin{aligned} F_{n\downarrow}(s) &= 1 - \alpha + \alpha s\downarrow \\ F_{n\uparrow}(s) &= 1 - \alpha + \alpha s\uparrow \end{aligned} \qquad n = 1,2,\ldots \tag{E.2}$$

where α is the probability that during the chain growth two monomers have reacted and the vector of auxiliary variables is s $(s\downarrow, s\uparrow)$.

The path-weight generating function is obtained in the same way as in the case of the AB_f polycondensates.

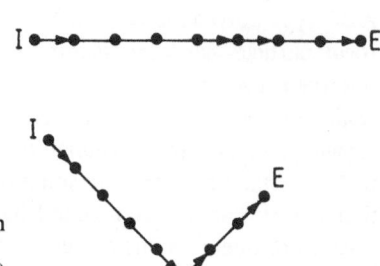

Fig. 56. Directed graph of a rooted tree from a linear chain polymerized by a chain reaction (free radical polymerization). *I* denotes the initiator, *E* the terminated chain end[106]

$$U_0(s) = s^{\phi_0}(1 - \alpha + \alpha U_{1\downarrow})(1 - \alpha + \alpha U_{1\uparrow})$$

$$U_{n\downarrow}(s) = s^{\phi_1}(1 - \alpha + \alpha U_{n+1,\downarrow}) \tag{E.3}$$

$$U_n\uparrow(s) = s^{\phi_1}(1 - \alpha + \alpha U_{n+1,\uparrow})$$

For the average of this path-weight generating function, one arrives after differentiation at

$$P_w\langle\phi\rangle_z = U_0'(1) = (1 + \alpha\phi)/(1 - \alpha\phi) \tag{E.4}$$

Again, we have made use of $\phi_n = \phi^n$ which holds for Gaussian chains. Equation (E.4) is formally identical to the equation for linear polycondensates. It differs, however, essentially in the meaning of α that is no longer the overall extent of reaction, which would be proportional to the monomer consumption, but is instead a conditional probability that an activated monomer has formed a bond with another monomer. This probability is only weakly dependent on the monomer conversion and cannot be determined by titration but has to be determined from kinetic measurements[107].

Branching due to the chain transfer of a side group of a polymer is easily described in the cascade mechanism by introducing a third functional group which may react with another monomer with probability γ. Instead of Eq. (E.1), one now has[106, 107]

$$F_0(s) = (1 - \alpha + \alpha s\downarrow)(1 - \alpha + \alpha s\uparrow)(1 - \gamma + \gamma s\uparrow) \tag{E.5}$$

all other steps remain the same as indicated by Eqs. (E.2) and (E.3) with the only difference that $U_{n\downarrow}(s)$ and $U_{n\uparrow}(s)$ have to be multiplied by the additional factor $(1 - \gamma + \gamma U_{n+1\uparrow})$. It has already been pointed out by Flory[222] that this kind of branching is not sufficient to reach the critical branching of gel-formation. The reason for this consists in a coupling of the two link probabilities α and γ as a result of a chain reaction. With increasing chain transfer, the chain length of the primary chain and thus α becomes smaller, and the sum of both effects, the chain growth and branching, never exceeds a critical value. The situation is changed drastically if the chain transfer due to the monomer is taken into account[106, 107]. The cascade theory predicts a gelation at rather high monomer conversion, and the gel formation has been in fact observed with PVAc in emulsion polymerization[168, 188, 189]. For details of the calculations and the use of kinetic data for the calculation of the link probabilities α and γ, we refer to the original papers. Here we wish to emphazise only one point:

The equations for the branching due to chain transfer at first sight resembles the ABC-type polycondensation. We might, therefore, expect a strong upturn of the reciprocal particle-scattering factor in a plot against q^2. This behavior is not, however, observed, nor predicted by theory. The reason again results from the peculiarity of a chain transfer reaction. On each chain transfer with a side group of the polymer, the growing linear chain is terminated and a new chain grows as a side chain from the created radical of the polymer. The situation is depicted in Fig. 57. Hence, each branch formation is necessarily accompanied by the formation of one free linear chain. The mass fraction of linear chains can easily be calculated from the cascade mechanism and is so large that the scattering behavior is essentially governed by the linear chain behavior[107].

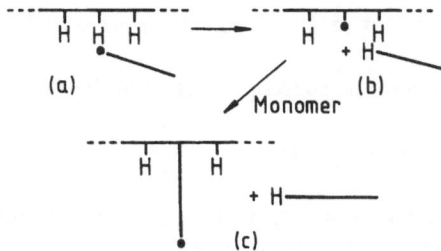

Fig. 57. Chain transfer in free radical polymerization with a side group of the "dead" polymer. Note: After termination of the radical one side chain *and* a free linear chain are obtained.

Linear chains obeying the most probable distribution do not show an upturn in the Zimm-plot[34, 156].

In a similar manner, the structure formation of polymer in free radical polymerization in the presence of a tetrafunctional chain transfer reagent was calculated[116]. Theory predicts the simultaneous formation of star-like molecules and linear chains, and to some extent cross-linking of the star-molecules as a result of radical recombination. Experiments gave reasonably good agreement with theory, as long as the chains were short. For large chains, the number of the chains attached to the tetrafunctional star centre scarcely increased beyond two on average. Evidently, the functional groups of the transfer active groups became widely shielded by a process not yet fully understood. Some suggestions for interpretation are discussed in the original papers.

A third example which has been studied in detail is the radical polymerization of divinyl monomers[117] which again was treated first by Gordon[31]. As in all chain reactions a primary chain is formed first, its length is determined by the link probability α, or the weight average degree of polymerization $P_{wp} = (1 + \alpha)/(1 - \alpha)$ and corresponds to the number of double bonds linked together by a growing radical in its activated state between initiation and termination. Each of the divinyl-monomers in this primary chain bears a pendant double bond, and an increasing fraction of them will be built in another primary chain, thereby forming a cross-link with further consumption of the monomer. Thus, the generating functions of the zero-th and higher generations can be written as[117]

$$F_0(s) = (1 - \alpha + \alpha s)^2 (1 - \gamma + \gamma (1 - \alpha + \alpha s)^2) \tag{E.6}$$

$$F_n(s) = (1 - \alpha + \alpha s) (1 - \gamma + \gamma (1 - \alpha + \alpha s)^2) \tag{E.7}$$

where γ is the probability that a pendant double bond was activated. The cascade substitution for the construction of the path-weight generating function is carried out as usual. After differentiation one finally finds the following relationships for the various averages[117].

$$P_w = P_{wp}/(1 - (P_{wp} - 1)\gamma) \cong P_{wp}(1 - \gamma/\gamma_c) \tag{E.8}$$

$$P_w P_z(q) = P_{wp} P_{zp}(q)/[1 - \gamma(P_{wp} P_{zp}(q) - 1)]$$

or

$$P_z(q) = \frac{(1 - \gamma/\gamma_c) P_{zp}(q)}{1 - (\gamma/\gamma_c) P_{zp}(q)} \tag{E.9}$$

$$\langle S^2 \rangle_z = \langle S^2 \rangle_{zp} (1 + \gamma)/(1 - \gamma/\gamma_c) \cong \langle S^2 \rangle_{zp}(P_w/P_{wp}) \tag{E.10}$$

where the subscript p refers to the properties of the primary chain and

$$\gamma_c = 1/(P_{wp} - 1) \tag{E.11}$$

is the critical cross-linking extent, where gelation takes place. The properties of the primary chains are given by

$$P_{wp} \quad = (1 + \alpha)/(1 - \alpha) \tag{E.12}$$

$$\langle S^2 \rangle_{zp} = (b^2/2)\alpha/(1 - \alpha^2) \tag{E.13}$$

$$P_z(q) \quad = (1 + \alpha\phi)/(1 - \phi\alpha) \cong (1 + 1/3 \langle S^2 \rangle_{zp} q^2) \tag{E.14}$$

with

$$\phi = \exp(- b^2 q^2/6)$$

Equation (E.8) is Stockmayer's well known result[7] and the other relationships were derived later by Whitney and Burchard[117]. Combination of Eqs. (E.14), (E.10) and (E.9) allows the particle-scattering factor of the cross-linked products to be rewritten as[117]

$$P_z(q) = (1 + 1/3 \langle S^2 \rangle_z q^2)^{-1} \tag{E.9'}$$

which is the relationship for the randomly branched polycondensates. Straight lines in the Zimm-plot of the copolymerization from MMA/DMMA have in fact been observed[117].

One easily verifies that the Eqs. (E.8) to (E.10) reduce to the relationships for the random polycondensates if the primary chain consists of one unit in length only, i.e. if the chain reaction becomes a step reaction. The factorized equations suggest that in many cases the structural properties of a primary molecule can be calculated separately, disregarding first the cross-linking reaction and inserting this result in an equation which takes account only of the cross-linking or branching reaction among the primary molecules. We shall come back to this point in chapter E.III, where we show that this conjecture is indeed often correct.

II. Random Vulcanization of Preformed Chains

In most cases, cross-linking is produced by vulcanization, where preformed primary chains of a certain length distribution are provided and where cross-links are introduced at random either by chemical reagents or by physical means, as for instance γ-ray irradiation. The general case was treated by Kajiwara and Gordon[119, 159], and for a full description either the first three moments of the molecular weight distribution is needed, or the z-averages of the mean-square radius of gyration, and the particle-scattering factor of the primary chains.

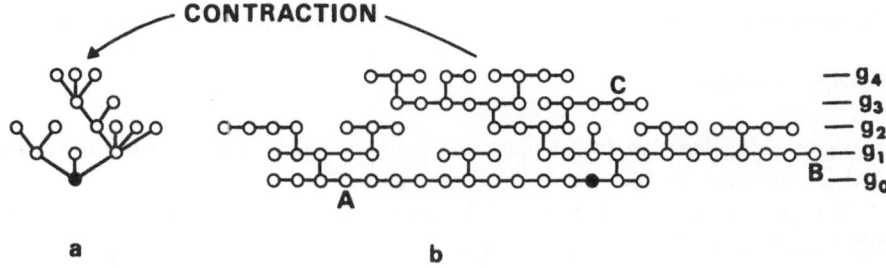

Fig. 58 a, b. Cross-linked tree (**b**) and its contracted form (**a**) with the root being marked by ●. A, B and C specify paths of length 10 from the root to the zero-th, first and third generations, respectively[159)]

The principles of the calculation by means of the cascade theory is sketched in Fig. 58 and compared with the random polycondensation. Instead of selecting a single mono-meric unit as root of a tree, a whole primary chain is placed on the zero-th generation, and the same is done for all the other primary chains from the cross-linked polymer.

The formalism of construction a path-weight generating function appears at first sight the same as for the random polycondensates, with the essential difference, however, that s and U_n in the equations for a polycondensate with functionality y for the monomers

$$U_0 = s^{\phi_0} (1 - \gamma + \gamma U_1)^y \tag{E.15 a}$$

$$U_1 = s^{\phi_1} (- \gamma + \gamma U_2)^{y-1} \tag{E.15 b}$$

have to be replaced by probability-generating functions for the primary chains, i.e.

$$s \quad \rightarrow \quad \sum_y m \, s^{B_0 \, (\phi, y)} \tag{E.16 a}$$

$$U_0 \quad \rightarrow \quad \sum_y m_y s^{b_0 \, (\phi, y)} (1 - \gamma + \gamma U_1)^y \tag{E.16 b}$$

$$U_1 \quad \rightarrow \quad \sum_y m_y s^{B_1 \, (\phi, y)} (1 - \gamma + \gamma U_2)^{y-1} \tag{E.16 c}$$

where m_y is the weight fraction of a primary chain of the degree of polymerization y, and the functions $B_j (\phi, y)$ replace the single weighting function ϕ_n in the simple cascade process. The weights $B_j (\phi, y)$ are actually generating functions, such that the coefficient of ϕ^n in the *Taylor expansion* of $B_j (\phi, y)$ gives the mean number of paths of length n which starts from a selected unit of the primary chain in the zero-th generation (root), and which ends in a unit on the j-th primary chain generation. Kajiwara and Gordon showed[159)] that $B_j (\phi, y)$ can be expressed in terms of $B_0 (\phi, y)$ as follows

$$B_j (\phi, y) = \phi^j y^{-1} [B_0 (\phi, y)]^2 \left[\frac{B_0 (\phi, y)^{-1}}{y - 1} \right]^{j-2} \tag{E.17}$$

and second that $B_0 (\phi, y)$ is given by

$$B_0 (\phi, y) = \frac{1 + \phi}{1 - \phi} - \frac{2}{y} \frac{(1 - \phi^y)}{(1 - \phi)^2} \tag{E.18}$$

Comparison of Eq. (E.18) with Eq. (C.12) discloses that for $\phi = \exp(-b^2q^2/6)$

$$B_0(\phi, y) = yP_y(q) \tag{E.19}$$

The generating function for enumerating paths of length n is actually the scattering function of isolated primary chains of DP = y.

The overall scattering function of the cross-linked system is obtained, as usual, by differentiation of U_0 at $s = 1$, which with Eq. (E.19) yields

$$P_wP_z(q) = \frac{(1 + \gamma\phi)\, P_{wp}P_{zp}(q)}{1 - [P_{wp}P_{zp}(q) - 1]\gamma} \tag{E.20}$$

For $\phi \simeq 1$ and $\gamma \ll 1$ this equation becomes identical to Eq. (E.8) for the cross-linking polymerization. The slight differences arises from the fact that Kajiwara and Gordon count the cross-links in the vulcanization process as bonds of finite dimension while in the special case of the chain reaction with divinyl monomers the double were assumed to be fixed at the same point, i.e. the cross-link is assumed dimensionless. Equation (E.9) to (E.14) also remain valid for the general vulcanization. The interesting part in the generalized derivation by Kajiwara and Gordon is that now for large primary chains the mean-square radius of gyration can be expressed by the z- and weight average DP of the primary chain:

$$\langle S^2 \rangle_z \simeq (b^2/6)\, (P_{zp}/P_{wp}) \tag{E.21}$$

We close this section with a few remarks. First

$$P_w/P_{wp} \equiv X_w = (1 + \gamma)/(1 - \gamma/\gamma_c) \tag{E.22}$$

is the weight average of cross-linked chains which is easily verified by setting $q^2 = 0$ in Eq. (E.20). Second, the translational diffusion coefficient and the angular dependence of the first cumulant can be obtained by integration over the particle-scattering factor as was indicated in Eq. (C.47) and (C.48)[84]. This yields for monodisperse primary chains and those with a most probable distribution at large X_w

$$b\langle R_h^{-1} \rangle_z \cong 3.685/P_w^{1/2} \tag{E.23 a}$$

(monodisperse primary chains)

$$b\langle R_h^{-1} \rangle_z \cong 3.464/P_w^{1/2} \tag{E.23 b}$$

(most probable distribution)

and independently of X_w:

$$\varrho = 1.504 \text{ for monodisperse primary chains} \tag{E.24 a}$$

$$\varrho = 1.732 \text{ for most probable primary chain distribution} \tag{E.24 b}$$

The angular dependence of Γ/q^2 has also been calculated[84]; it appears not to differ seriously from those of randomly cross-linked or randomly branched polymers. Furthermore, the ϱ-parameter of monodisperse fractions obtained by Lagrange expansion, has been calculated and is listed in the original paper[84]. The parameter decreases from 1.504 for $X = 1$, i.e. no cross-linking, to 1.1318 for very large molecular weights, i.e. $X \gg 1$. These calculations confirm the observation made before that ϱ decreases with increasing uniformity.

III. Heterogeneities[95]

The term heterogeneity entails a rather wide conception. Generally speaking, heterogeneity is defined by a more or less sudden change in a property when passing from one point in space to another. So we may have
(i)   heterogeneities in optical density or in the refractive index (Schlieren) or
(ii) heterogeneities in density or branching density and finally
(iii) heterogeneity in flexibility resulting for instance in different glass transition points.
All these types of heterogeneities can, at least in part, be treated by the cascade theory. The first case, i.e. optical inhomogeneities have been considered, in passing, when general copolymers were treated. Nothing was said, however, on the extension and the structure of the domains. For instance, the domains may be distributed in space at random and may have a random size distribution. This case is observed with random copolymers. At the other extreme, domains with rather well defined surfaces may be embedded in a more or less homogeneous matrix of interconnecting chains. This extreme corresponds to what is named by physicists the "two phase model". The real cases will be found somewhere in between these two extremes. We shall here discuss two approaches in brief. The first consists of the coupling of preformed domains through linear chains and the second is a special inhomogeneity due to a substitution effect. In the former case long range correlations are introduced between *different generations*, in the latter the correlation is introduced between units in the *same generation*. Two examples of a substitution effect are sketched in Fig. 59. We speak of a first shell or second shell etc. substitution effect depending on the number of generations which we have to retrace before the paths of two units concerned meet in one common unit.

Fig. 59. a First shell substitution effect: the probability for a reaction of functionality x depends on whether y or z or both had already reacted. **b** Second shell substitution effect: the probability of reaction for a unit x_1 and x_2 depends on how many of the functional groups in y and z had reacted

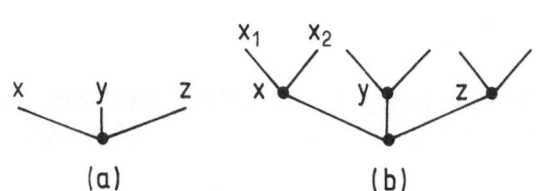

1. Coupling of Domains[95, 135)]

The problem may be exemplified by two cases:
(i) where molecules of an A_f polycondensate are coupled by linear chains of the B_2-type polycondensate
(ii) where the nuclei are AB_2 polycondensates coupled via linear chains of the CD-type.

a) Coupling of A_f with B_2 Domains

The individual domains indicated by superscripts 0, are defined through two sets of generating functions for the various generations

A_f domain (branched):

$$F_{0A}^0(s) = (1 - \alpha + \alpha s_A)^f$$

$$F_{NA}^0(s) = (1 - \alpha + \alpha s_A)^{f-1}$$

(E.25)

B_2 domain (linear):

$$F_{0B}^0(s) = (1 - \beta + \beta s_B)^2$$

$$F_{nB}^0(s) = (1 - \beta + \beta s_B)$$

(E.26)

Passing through the stage of cascade substitution for the construction of the path-weight generating functions, one finally arrives at

$$U_{0A}^{0'}(1) = 1 + \alpha f \sum_{n=1}^{\infty} \alpha(f-1)^{n-1}\phi_{nA} = 1 + \alpha f \phi_A^* = [P_w P_z(q)]_A \tag{E.27}$$

$$U_{0B}^{0'}(1) = 1 + \beta 2 \sum_{m=1}^{\infty} \beta^{m-1}\phi_{mB} = 1 + 2\beta\phi_B^* = [P_w P_z(q)]_B \tag{E.28}$$

For reasons which will become clear later on, we have introduced the abbreviations

$$\phi_A^* = \sum_{n=1}^{\infty} [\alpha(f-1)]^{n-1}\phi_{nA} \tag{E.29 a}$$

$$\phi_B^* = \sum_{m=1}^{\infty} [\beta^{m-1}]\phi_{mB} \tag{E.29 b}$$

These functions ϕ^* represent the scattering behavior of a full subtree that emerges from a functional group of a monomer selected at random. For instance:

$$\phi_A^* = [P_w P_z(q) - 1]/\alpha f \xrightarrow{\text{large } P_w} P_w P_z(q)/\langle f \rangle \tag{E.30}$$

where $\langle f \rangle$ denotes the average number of functional groups which have reacted.

Coupling of the two domains is achieved by partial reaction of the free end-groups of the two polymeric blocks. The fraction of free A and B groups is $(1 - \alpha)$ and $(1 - \beta)$, respectively, and $(1 - \alpha)p$ and $(1 - \beta)q$ are the fraction of these free endgroups which may be consumed by the coupling. We thus obtain for the zero-th generations

$$F_{0A}(s) = [(1 - \alpha)(1 - p) + (1 - \alpha)ps_B + \alpha s_A]^f \tag{E.31}$$

$$F_{0B}(s) = [(1 - \beta)(1 - q) + (1 - \beta)qs_A + \beta s_B]^2 \tag{E.32}$$

and similar expressions for the other generations. Performing the cascade substitution as usual for the copolymers, one finally obtains

$$U'_{0A} = \phi_{0A} + \alpha f U'_{1A} + p(1 - \alpha)f U'_{1B}$$
$$U'_{0B} = \phi_{0B} + 2\beta U'_{1B} + 2(1 - \beta)q U'_{1A} \tag{E.33}$$

$$U'_{1A} = \phi_{1A} + \alpha(f - 1)U'_{2A} + (1 - \alpha)(1 - f)p U'_{2B}$$
$$U'_{1B} = \phi_{1B} + (1 - \beta)q U'_{2A} + \beta U'_{2B} \tag{E.34}$$

This set of equation can be factorised as was shown in Chap. C, as

$$P_w P_z(q)_{app} = \hat{m} \cdot [\phi_0 + \langle N(1)\rangle \phi(1 - P\phi)^{-1}] \cdot \hat{M} . \tag{E.35}$$

if Gaussian statistics is assumed, where

$$\phi = \begin{bmatrix} \phi_A & 0 \\ 0 & \phi_B \end{bmatrix} \qquad\qquad \hat{M}_0 = \begin{bmatrix} \nu_A M_{0A} \\ \nu_B M_{0B} \end{bmatrix}$$

and $\quad \hat{m} = (n_A \nu_A M_{0A}, n_B \nu_B M_{0B}) \tag{E.36}$

$\langle N(1)\rangle$ and P are the coefficient matrices in Eqs. (E.33) and (E.34)

Equation (E.35) is very convenient for computational work but has two disadvantages. First, the equation is restricted to Gaussian behavior of the subchains, and second, the block-like structure is not immediately recognizable. The block character of the copolymer can, however, be made evident by a different factorization. We first realize that Eq. (E.34) can be written as

$$U'_{1A} = \sum_{n=1}^{\infty} [\alpha(f - 1)]^{n-1}\phi_{nA} + (f - 1)(1 - \alpha)p \sum_{n=1}^{\infty} [\alpha(f - 1)]^{n-1}U'_{n+1, B} \tag{E.37}$$

$$U_{1B} = \sum_{m=1}^{\infty} \beta^{m-1}\phi_{mB} + (1 - \beta)q \sum_{m=1}^{\infty} \beta^{m-1}U'_{m+1, A}$$

Next, we assume flexible links between the two types of blocks. Then a path-weight function $\phi_{n+m, B}$ starts with a unit from the A-block, passes through n units of this block, crosses over to the B-block, and ends after m B-units with a B unit. Therefore, we can write

$$\phi_{n+m,B} = \phi_n^{(A)} \cdot \phi_m^{(B)} \tag{E.38}$$

where the superscripts denote pure A or B blocks, respectively. With this assumption, it can further be shown that

$$U'_{n+1,B} = \phi_n^A \cdot U'_{1B}$$
$$U_{m+1,A} = \phi_m^B \cdot U'_{1A} \tag{E.39}$$

and

$$U'_{1A} = \phi_A^* + (f-1)(1-\alpha)\,p\,\phi_A^*\,U'_{1B}$$
$$U'_{1B} = \phi_B^* + (1-\beta)\,q\,\phi_B^*\,U'_{1A} \tag{E.40}$$

which can be factorized in matrix notation as

$$U'_1 = \phi^* (1 - \phi^* P^*)^{-1} \tag{E.41}$$

where

$$\phi^* = \begin{bmatrix} \phi_A^* & 0 \\ 0 & \phi_B^* \end{bmatrix} \tag{E.42}$$

and

$$P^* = \begin{bmatrix} 0 & (f-1)(1-\alpha)\,p \\ (1-\beta)\,q & 0 \end{bmatrix} \tag{E.43}$$

where the ϕ_j^* are given by Eq. (E.29 a) and (E.29 b). Finally, when Eq. (E.41) is combined with Eq. (E.33)

$$P_w P_z(q)_{app} = \hat{m} \cdot [\phi_0 + N(1)\,\phi(1 - \phi^* P^*)^{-1}]\hat{M}_0 \tag{E.44}$$

Comparison with Eq. (E.35) shows a close similarity but with the relevant difference that the matrix ϕ^* is multiplied on the left with the reduced transition-probability matrix P^* while in Eq. (E.35) ϕ it is multiplied with P on the right[223].

The physical meaning of Eq. (E.44) is essentially different from that of Eq. (E.35) since the diagonal elements of ϕ^* contain now the contributions of a whole homoblock instead of only one unit as in the case of Eq. (E.35). Furthermore, the elements of P^* contain only the transition probabilities from one block to the next. All the other transition probabilities within a homoblock are included in P^*. Figure 60a demonstrates in a graph the fully displayed rooted tree while Fig. 60b shows the contracted tree which

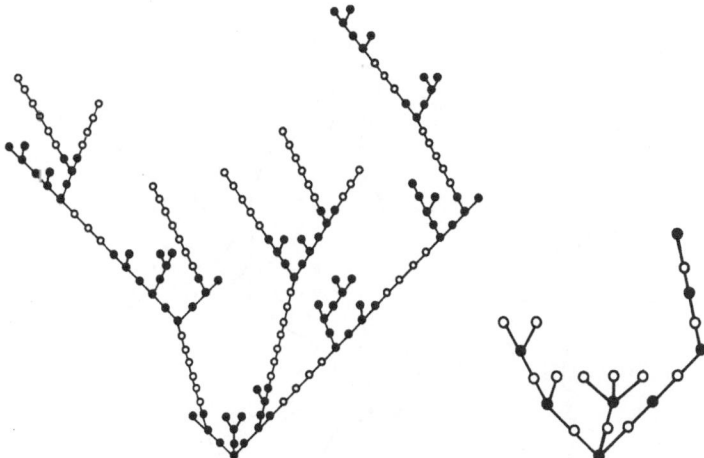

Fig. 60. Representation of a branched block-copolymer as a full rooted tree (*left*) and as a reduced rooted tree (*right*)[95]

corresponds to the transition matrix **P***. The white and black circles represent blocks. The graphs of Fig. 60, however, demonstrate correctly only the conditions for the transition matrices; for the population in the first generation, in both cases the non-reduced matrices $\langle \mathbf{N}(1) \rangle$ are needed[95].

Equation (E.42) to (E.44) have the great advantage that all properties of the homo-block can be calculated separately, and the properties of the branched block-copolymer are obtained by using the reduced transition matrix **P*** which contains only the coupling probabilities. Note that these coupling appear only as off-diagonal elements.

b) Coupling of AB₂ and Linear CD Blocks

In heterogeneously branched glycogen, blocks of the AB_2-type polycondensates appear to be linked via a linear chain of the CD-type polycondensate[95, 135]. Figures 61 and 62 show details of the structure in different scales of magnification. This coupling reaction is slightly more complicated because of the constraints for the reaction between the unlike functional groups A and B, and C and D, respectively. The final equations can be derived in a similar manner with the same formal solution as given by Eq. (E.44) but where the two matrices $\langle \mathbf{N}(1) \rangle$ and **P*** have the elements

$$N(1) = \begin{bmatrix} 2\beta & \alpha & 2(1-\beta)\,p_c & (1-\alpha)\,p_D \\ (1-\gamma)^w A & (\gamma)^w B & \gamma & \gamma \end{bmatrix} \tag{E.45}$$

$$P^* = \begin{bmatrix} 0 & 0 & 2(1-\beta)\,P_c & 0 \\ \beta & 0 & (1-\beta)\,P_c & (1-\alpha)\,P_D \\ (1-\gamma)^w A & 0 & 0 & 0 \\ 0 & (1-\gamma)\,w_B & 0 & 0 \end{bmatrix} \tag{E.46}$$

and the matrix ϕ^* now has 4 diagonal elements ϕ_b^*, ϕ_b^*, ϕ_l^*, ϕ_l^*, where the indices b and l refer to the branched and linear domains, respectively

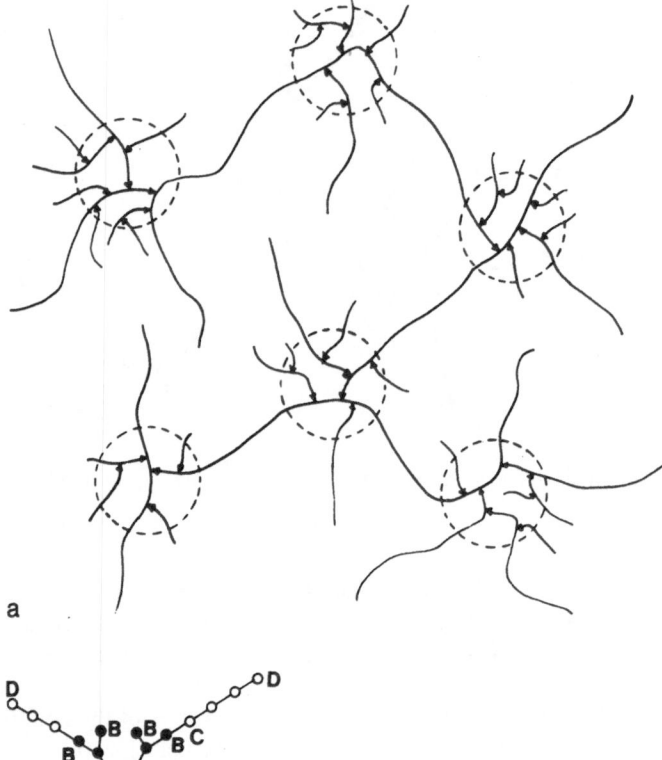

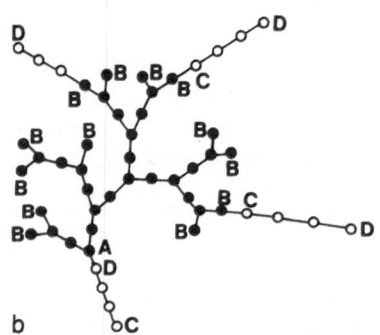

a

b

Fig. 61. a Model of a heterogeneously branched macromolecule and **b** a detail of the strucutre[95)]

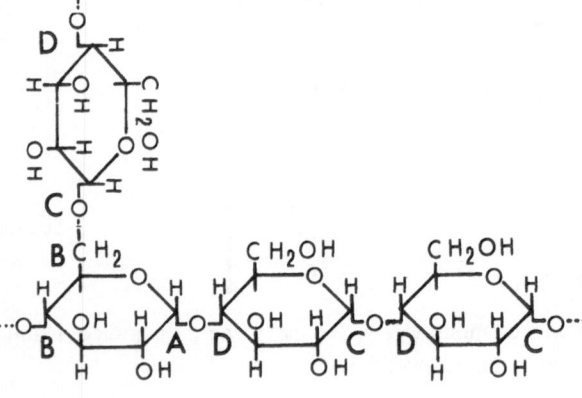

Fig. 62. Chemical structure of a section of a glycogen molecule. The letters A, B, C, D denote the functional groups[95)]

$$\phi_b^* = \sum_{n=1}^{\infty} \alpha^{n-1} \phi_b^n \, ; \qquad \phi_1^* = \sum_{m=1}^{\infty} \gamma^{m-1} \phi_{1m} \qquad\qquad (E.47)$$

with the definitions as given by Eq. (E.29). The probabilities α and γ denote the reaction of an A group with a B group and of a C group with a D group in the linear chain, respectively. Because of the constraint, we have $\alpha = 2\beta$. The **P*** matrix now contains an off-diagonal element in position p_{21}, and another one in position P_{23} which denotes the mentioned constraint between the A and B groups.

c) Heterogeneity in Flexibility[224]

One special type of heterogeneity arises from interconnecting chains which are rigid rods jointed by freely moving hinges[225]. Then the path-weight is given by

$$\phi_{1m} = \frac{\sin(bmq)}{bmq} \qquad\qquad (E.48)$$

where $bm = l_m$ is the fibre (rod) length between the branched domains. Evidently, in this case $\phi_{1m} \neq (\phi_1)^m$, and the sum

$$\phi_1^* = \sum_{m=1}^{\infty} \gamma^{m-1} \frac{\sin(bmq)}{bmq} \qquad\qquad (E.47\,a)$$

can no longer be solved analytically. However, since $\gamma < 1$, the sum will converge fairly quickly, and ϕ_1^* can easily be obtained numerically on a computer to any desired accuracy.

This technique was applied for a description of the scattering behavior of branched fibrils from fibrin in the molecular weight region, where first only a rod growth was observed, up to the point of clotting, i.e. gel formation. Since electron microscopic pictures (EM) showed only single branching points, the AB_2-domains appeared to consist of one unit only. This implies

$$\alpha = \beta = 0 \qquad\qquad (E.49\,a)$$

For reasons which are discussed in the original paper, we could set

$$p_C = w_A = w_B \qquad\qquad (E.49\,b)$$

Finally, the condition of the basic theorem Eq. (C.114) requires

$$2(1-\beta)\,p_C w_A = (1-\alpha)\,p_D w_B$$

or with Eq. (E.49 a) and (E.49 b)

$$2\,p_C = p_D \qquad\qquad (E.49\,c)$$

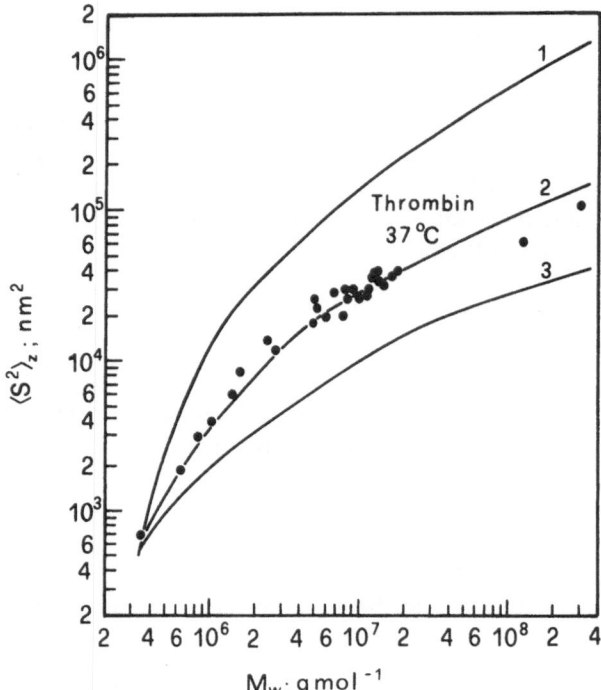

Fig. 63. Dependence of $\langle S^2 \rangle_z$ on M_w for fibrin initiated by thrombin at 37 °C (−●−); *curve 1* gives the theoretical dependence for a single strand, *curve 3* shows the behavior of fibres with many strands parallel, and *curve 2* is the best fit[224)]

The matrices $\langle N(1) \rangle$ and P* in Eq. (E.45) and (E.46) reduce to simple forms and may be evaluated even by hand. This was done in the original paper.

The conditions (E.49) leave us with only two adjustable parameters γ and p_C. γ defines the fibre length between two branching points by the relationship $P_{wl} = (1 + \gamma)(1 - \gamma)$, and p_C the overall molecular weight M_w. Figure 63 shows the observed and the calculated M_w dependence of $\langle S^2 \rangle_z$ for fibrin. Very good agreement could be achieved by introducing a certain thickness of the fibres, i.e. several strands are packed parallel. The theory allowed the prediction of a fibre length of 530 nm which agrees almost perfectly with the EM obervations, where 525 nm was found from fibrin networks. Here also a higher fibre thickness than that of a single strand was found.

With the probabilities γ and p_C, once determined, the scattering curves are fully determined and, in fact, the calculations describe the scattering curves in almost all details. In particular, the striking change in behavior at the onset of branching is correctly described. This onset of branching can also be recognized from Fig. 63; the steep increase in $\langle S^2 \rangle_z$ with the molecular weight displays the linear fibre growth but the flat continuation beyond $M_w = 4 \cdot 10^6$ is significant for branching.

2. Prevention of Segment Overcrowding Treated as a Second Shell Substitution Effect[177)]

This problem was touched in Chap. D.I.3 when the critical exponents were briefly discussed. Here we discuss the special case of a steric hindrance, for instance to avoid units packing too close. The model treated by Gordon and Parker assumes that in a three functional polycondensate two units which are directly linked, cannot have all of their remaining four functionalities converted to links: at least *one* must remain unreacted. Figure 64 demonstrates by the graph of a rooted tree, which structure will occur, even at the highest possible extent of reaction.

To meet the conditions of steric hindrance, one may be guided by the graph of Fig. 64 and tentatively assume

(i) a unit selected at random can react at random with all of its three functionalities yielding in the zero-th generation $F_0^0(s) = (1 - \gamma + \gamma s)^3$

(ii) The reactivity in the first generation depends on how many units were already linked to the other functionalities of the same generation. This fact can be approximately taken into account by giving the occupation of the first generation different labels: if only one functionality of the root had reacted, the generating function for the first generation might be labelled $F_{11}^0 = (1 - \gamma + \gamma s)^2$, where the first index refers to the generation number while the second index gives the label. If two of the three functionalities had reacted, the label might be $F_{12}^0 = F_{11}^0$, and finally, if all three functional groups have reacted, we insist that only one of the two remaining functionalities could react, i.e. $F_{13}^0 = 1 - \gamma + \gamma s$. (The superscript 0 should indicate a zero-th approximation which will be improved in the following:)

We have now to take care that the various generations are properly connected. This is achieved as usual by introducing labels s_1, s_2, and s_3 to the auxiliary variables. Furthermore, there may be reasons to assume that there will be a certain hindrance in the reaction if adjacent functionalities are already occupied. (First shell substitution effect.) Taking all these requirements together, Gordon and Parker were led to set up the generating functions as follows:

$$F_0(s) = c_0[(1 - \gamma)^3 + 3(1 - \gamma)^2(\gamma/c_1) s_1 + 3(1 - \gamma)(\gamma/c_2)^2 s^2 + (\gamma/c_3)^3 s_3^3]$$

$$F_{12}(s) = c_0 c_2[(1 - \gamma)^2 + 2(1 - \gamma)(\gamma/c_2) s_1 + (\gamma/c_3)^2 s_2^2]$$

$$F_{11}(s) = F_{12}(s)$$

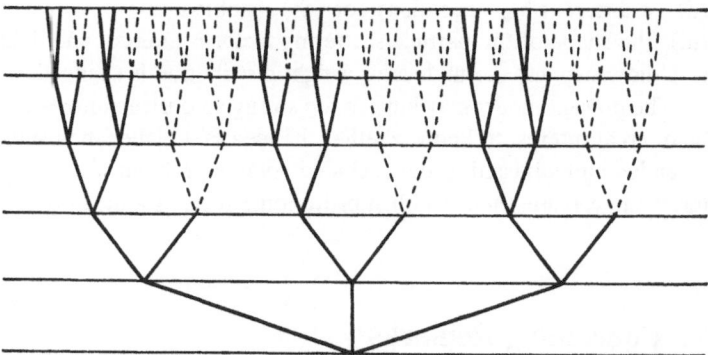

Fig. 64. Effect of the steric hindrance condition on the equilibrium gel at maximum attainable conversion: the dotted parts are removed by "switching on" the effect of Eq. (E.50). Only generation from 0 to 5 are shown[177)]

$$F_{13}(s) = c_0 c_3 [(1 - \gamma)^2 + 2(1 - \gamma)(\gamma/c_2) s_1]$$

$$F_{i1}(s) = F_{11}(s) = F_{12}(s) \qquad\qquad\qquad\qquad\qquad\qquad (E.50)$$

$$F_{i2}(s) = F_{13}(s)$$

The factor $c_1 = c_2$ and c_3 introduce the influence of the neighboured functionalities but we wish to make sure that a reaction of only two functional groups is not inhibited. One verifies from a substitution $F_0(F_1(s))$ that this is indeed achieved by the special choice of the factors c_2 and c_3 in Eq. (E.50). The three constants c_0, c_2 and c_3 have to be determined such that they fulfil the normalization condition for the probability-generating functions. This means a solution of the three coupled equations

$$c_0 [(1 - \gamma)^3 + 3(1 - \gamma)^2 (\gamma/c_2) + 3(1 - \gamma)(\gamma/c_2)^2 + (\gamma/c_3)^3] = 1$$

$$c_0 c_2 [(1 - \gamma)^2 + 2(1 - \gamma)(\gamma/c_2) + (\gamma/c_3)^2] = 1 \qquad\qquad (E.51)$$

$$c_0 c_3 [(1 - \gamma)^2 + 2(1 - \gamma)(\gamma/c_2)] = 1$$

The solution of this system of equation leads to the recognition that the c_i depend on γ, and therefore γ is no longer a quantity which describes the extent of the reaction α. The measureable extent of reaction is obtained from $F_0(s)$ by differentiation at $s = 1$ as usual

$$F_0'(1) = 3 c_0 [(1 - \gamma)^2/c_2 + 2(1 - \gamma)\gamma/c_2^2 + \gamma^2/c_3^3] = 3\alpha$$

or   (E.52)

$$\alpha = c_0 [(1 - \gamma)^2/c_2 + 2(1 - \gamma)\gamma/c_2^2 + \gamma^2/c_3^3]$$

The rest of the work consists in the cascade substitution for the construction of a path-weight generating function which is done as usual. Gordon and Parker have focussed their interest on the position of the gel point which became shifted from $\alpha_c = 0.5000$, in the random case, to $\alpha_c = 0.5733$, for the steric hindrance case. It will certainly be of interest to investigate whether
(i)   the molecular weight dependence on $(1 - \alpha/\alpha_c) = \varepsilon$ is changed
(ii)   a different M_w dependence of $\langle S^2 \rangle_z$ is obtained
(iii) the shape of the particle scattering factor is altered, and finally
(iv)   how strongly the molecular weight distribution is changed.
        The present model may impose too strong an obstruction on a real system, and it will be of great interest to know whether the idea of a higher shell substitution effect can be extended and adjusted to the excluded volume problem. Certainly, nobody can characterize the cascade theory with substitution effects as a mean field theory.

F. Concluding Remarks

In the present article we have reviewed the utility and progress of the FS theory from the early beginning in the 1940's to the recent success in interpreting experimental results. It may be useful at the end to add some general remarks on the possibilities and applicability of the FS theory on the one hand, and their limitations on the other.

Viewing the literature on branched polymers, it is striking that even in recent research work interest is focussed mostly on predicting the molecular weights M_w and M_n, and the corresponding molecular weight distribution from the conditions of a chemical preparation. These quantities, and in particular the conditions for gelation, are of great interest and practical relevance. Nevertheless, this restriction surprises since we know it is the structure in space, the conformation, which determines the properties of macromolecules.

The renunciation of a calculation of these conformational properties appears needless and unreasonable both when the FS theory is applied in its versatile modification of the cascade branching theory, or when the percolation theory is used. In the cascade theory all conformational properties which can be expressed as a sum of functions ϕ (j − k), are economically obtained from one path-weight generating function by using appropriate weighting functions. In the percolation theory, on the other hand, the cluster size and its distribution are derived mostly by computer simulations on a lattice. Here again, all parameters needed for the calculation of conformational properties are already available. Still only the mean-square radius of gyration has been calculated so far. The lack of data for the diffusion coefficient, the particle-scattering factor, and the first cumulant is regrettable since both branching theories seem in a way to be complementary to each other, i.e. both have their limitations, but at different points.

This may be demonstrated with the problem of ring formation as an example. The possibilities of forming cycles and larger loops can at present only insufficiently be treated by the cascade theory. If the macromolecule contains no loops and the rooted tree representation is used, the repeating units appear uniquely to be placed on different generations. However, when loops are formed, two units from different generations are linked together, and the assignment to special generations becomes difficult and ambivalent. If only the molecular weight M_w is of interest, the influence of ring formation can to some extent satisfactorily be described by the spanning-tree approximation. Graph theory shows that each general graph that contains cycles can be transformed in a spanning tree by cutting one of the bonds in a cycle, whereupon the adjacent repeating units of a ring now appear on different generations. This approximation was first used by Gordon and Scantlebury[30], where they treated the two cut functionalities as being wasted, i.e. not being capable for a branching or cross-linking reaction. The spanning-tree approximation has been extensively applied with good success by Dušek and his coworkers[226–229], and we refer here to his papers for more details. Dušek and Gordon could calculate the shift of gel point to higher conversions of the functional groups and the change in the properties of the gel not too far beyond the gel point. In most cases, the extent of ring formation was found to be small.

The spanning tree approximation breaks down when conformational properties are to be calculated. This becomes clear from a consideration of the mean-square radius of gyration. For a linear chain that forms an end-to-end coupled ring, the mean-square radius of gyration decreases by a factor two, i.e. $\langle S^2 \rangle_{ring} = (1/2) \langle S^2 \rangle_{lin}$[13], and for a path in a branched molecule which contains n_r cycles, it has been estimated that $\langle S^2 \rangle_{path} = (1 + n_r)^{-1} \langle S^2 \rangle_{linear}$[230]. This shrinking of dimensions is neglected in the spanning-tree approximation. Moreover, if the Stockmayer-Jacobson[9] treatment of ring formation in linear chains is tentatively extended to branched polymers, theory predicts that most of the rings are small cycles. This prediction stands in contrast to experimental observations on branching reactions in dilute solutions. Under these condition, microgels are formed where functional groups in the surface area react extensively until all functional groups in this domain are consumed. Such reactions necessarily include the formation of large rings.

Percolation simulations on a computer on the other hand demonstrate nicely the formation of a well defined surface[43–45]. The prediction of the gel point still remains poor; probably the extent of ring formation is overestimated in the presently used simplest models for percolation.

Closely related to the question of ring formation is the problem of excluded volume. The thermodynamic interaction between segments decays strongly with their mutual distance and is noticeable only when the segments come into contact. This implies that temporarily loops must be formed. For weak interaction, perturbation theory can be applied and the excluded volume can be calculated by the cascade theory[34, 176]. The region of validity of such perturbation theory is small for most branched polymers since the segment density is here much higher than for linear chains. Some interesting suggestions have been made recently by Daoud and his coworkers[154] and by Lubensky and Isaacson. Daoud et al.[154, 231] have extended the "blob" conception to star-shaped molecules[154] and random f-functional polycondensates[231]. Apparently, the behavior of star molecules is reasonably described by this picture. Most interesting and promising is a recent theory by Lubensky and Isaacson[232] who developed on the basis of a field theory a generating function calculus, where the partition function of a real system is taken into account. Specific models have not been calculated by this method.

All excluded volume theories for branched chains suffer, however, from a principal deficiency since the assumption is tacitly made that all monomeric units in the molecule may have, in principle, the chance to interact with each other. This is, however, a too extensive assumption since for sterical reasons two remote segments can never form a contact. Hence, the excluded volume effect is highly overestimated for densely branched chains. In fact, highly branched polymers show the phenomenon of swelling but no detectable distortion of Gaussian chain behavior[117, 137, 138, 179].

Acknowledgement. The author received his basic knowledge on the Cascade theory from Professor Manfred Gordon, Essex University, England, who introduced him 1969 into this exciting field in his characteristically unconventional manner (while we were waiting for a dinner). I am deeply indebted to his kind patience and the many stimulations, which he gave in the course of the past decade. With gratitude I acknowledge the pleasing joint work with Professor Walter H. Stockmayer, Dartmouth College, Hanover, U.S.A. while he was in Freiburg, on the various aspects of the dynamic light scattering. I am grateful to Dr. Kanji Kajiwara, Kyoto University, who showed me so many tricks how to handle mathematical problems and whose frequent comments during his stay in Freiburg (1981–1982) were very helpful while the manuscript was in a stage of fermentation. This fermentation became appreciably accelerated by discussions with Dr. Simon B. Ross-Murphy, Unilever Research, Sharnbrook, England. He also helped me to bring the manuscript into a readable form. Finally, I wish to thank my students and coworkers who all picked up the merging problems with enthusiasm and who helped me with their candid questions to come to a deeper understanding of the topic. Mrs. Gerda Räpple took the pain of typing the manuscript. This article would be inconceivable without the patience and encouragements of my wife Else-Marie during a time of serious difficulties and despondency.

G. Appendix

Lagrange Expansion

A probability-generating function $W(s)$ is defined by

$$W(s) = \sum w_x s^x \tag{A1}$$

where w_x is the corresponding probability distribution. Thus, this distribution can be obtained from $W(s)$ by a *Taylor expansion*, where the coefficients of s^x form the desired probability distribution.

Often, however, $W(s)$ is not known as an explicit function of s, but only implicitly as function of, say, $u(s)$. Lagrange developed in 1770 a technique, where the desired coefficient of s^x is obtained by an expansion of $W(u(s))$ in terms of $u(s)$. Let

$$W(s) = s F_0(u(s)) \tag{A2}$$

with

$$u(s) = s F_1(u) \tag{A3}$$

then according to Lagrange, the distribution is given by

$$w_x = \frac{1}{(x-1)!} \frac{d^{x-2}}{d_u^{x-2}} [F_0'(u) \cdot F_1(u)^{x-1}]_{u=0} \tag{A4}$$

where

$$F_0'(u) = d F_0(u)/du$$

Example: Weight distribution of linear random polycondensate

The p.g.f. for the f-functional random polycondensate was defined in Chap. IV.3 by the g.f. $F_0(u)$ and $F_1(u)$. For the linear polycondensates, we have $f = 2$. Hence

$$F_0(u) = (1 - \alpha + \alpha u)^2 \tag{A5}$$

$$F_1(u) = 1 - \alpha + \alpha u$$
$$F_0'(u) = 2\alpha(1 - \alpha + \alpha u) = 2\alpha F_1(u) \tag{A6}$$

Therefore, the distribution is

$$w_x = \frac{2\alpha}{(x-1)!} \frac{d^{x-2}}{d_u^{x-2}}[(1 - \alpha + \alpha u)^x]_{u=0} \tag{A7}$$

Performing the $(x-2)$ fold differentiation, one obtains

$$w_x = 2\,\alpha^{x-1}\,(1-\alpha)^2\,[x\,(x-1)\ldots 3]/(x-1)! \tag{A 8}$$

$$w_x = (1-\alpha)^2\,x\,\alpha^{x-1}$$

which is in fact the well known weight fraction of the most probable distribution.

The *Lagrange expansion technique* can also be applied to the calculation of the particle-scattering factors $P_x\,(q)$ of branched or linear polymers of DP = x from the path-weight generating function of the polydisperse system. In Chap. C.III we have shown the equivalence

$$P_w \cdot P_z(q) = \sum w_x \left[x^{-1} \sum_{j}^{x} \sum_{n=0}^{\infty} N_j\,(n)\,\phi_n \right] \tag{C 23}$$

where j denotes a special unit from an x-mer that was chosen as root for a tree and $N_j\,(n)$ is the number of units in the n-th generation for this tree. Eq. (C 23) can also be written as

$$P_w P_z(q) = \sum w_x [x\,P_x(q)] \tag{A 9}$$

Comparison of Eq. (A 9) with (C 23) shows that $x\,P_x(q)$ could be derived from a generating function of the following kind

$$s_1 H\,(s,q) = \sum_{x=1}^{\infty} w_x [x\,P_x(q)]\,s_1^x \tag{A 10}$$

which differs from W (s) by the extra factor[4] $x\,P_w\,(q)$. Our aim is now to find a procedure which allows to generate this extra factor from relationships already known to us. We know already two properties of the path-weight generating function defined in Eq. (C 86)

$$U_0\,(s) = s^{\phi_0}\,F_0\,(u_1) \tag{C 86}$$

$$U_1\,(s) = s^{\phi_1}\,F_1\,(u_2)$$

etc.

These are

(1) $\quad \dfrac{d\,U_0\,(s)}{ds}\bigg|_{s=1} = \phi_0 + F_1'\,(1)\sum F_1'\,(1)^{n-1}\,\phi_n = \sum w_x x^{-1} P_x\,(q)$

if $\quad \phi_n = \exp\,(-b^2 q^2/6)^n$

and

(2) $\quad \dfrac{d\,U_0\,(s)}{ds}\bigg|_{s=1} = 1 + \dfrac{F_1'\,(1)}{1-F_1'\,(1)} = \sum w_x x$

if $\quad \phi_n = 1 \qquad$ i.e. $\quad q = 0$

We now make a guess, how Eq. (A 10) can be derived. Let us set

$$s^{\phi_n} = s_1 s_2^{\phi_n} \tag{A 11}$$

Eq. (C 86) takes then the form

$$U_0(s_1, s_2) = s_1 s_2^{\phi_0} F_0(U_1)$$
$$U_1(s_1, s_2) = s_1 s_2^{\phi_1} F_1(U_2) \tag{A 12}$$

etc.

This bivariate generating function has the two familiar limiting properties

$$\left.\frac{\partial U_0(s_1, 1)}{\partial s_1}\right|_{s_1 = 1} = \sum x w_x \tag{A 13}$$

$$\left.\frac{\partial U_0(1, s_2)}{\partial s_2}\right|_{s_2 = 1} = \sum w_x \, x \, P_x(q) \tag{A 14}$$

If we differentiate at $0 \leq s_1 \leq 1$ with respect to $s_2 = 1$, we obtain

$$\left.\frac{\partial U_0(s_1, s_2)}{\partial s_2}\right|_{s_2 = 1} = \sum w_x s_1^x \, x \, P_x(q) = s_1 H(s_1, q) \tag{A 15}$$

In fact, Eq. (A 14) is recovered for $s_1 = 1$ or $q = 0$. Thus, with Eq. (A 12) we find

$$H(s_1, q) = \phi_0 F_0(u) + \left[\frac{d F_0(u)}{du} \frac{d U_1}{d s_2}\right]_{s_2 = 0} \tag{A 16}$$

$$\left.\frac{d U_1}{d s_2}\right|_{s_2 = 1} = s_1 F_1(u) \sum \left(s_1 \frac{d F_1}{du}\right)^{n-1} \phi_n \tag{A 17}$$

and since

$$u = s_1 F_1(u)$$

we have

$$\left.\frac{d U_1}{d s_2}\right|_{s_2 = 1} = u \sum \left(\frac{u}{F_1(u)} \frac{d F_1}{du}\right)^{n-1} \phi_n \tag{A 18}$$

Thus,

$$H(s_1, q) = \phi_0 F_0(u) + u F_0'(u) \sum \left[\frac{u}{F_1(u)} F_1'(u)\right]^{n-1} \phi_n \tag{A 19}$$

with

$$u = s_1 F_1(u) \tag{A 3}$$

If we compare this result with Eq. (A 2), and Eq. (A 3) we notice that the function

$$s_1 H(s_1, q) = \sum w_x x P_x(q) s_1^x \tag{A 10}$$

can be expanded according to Lagrange's formula with the result

$$w_x x P_x(q) = \frac{1}{(x-1)!} \frac{d^{x-2}}{du^{x-2}} [H'(u) F_1(u)^{x-1}]_{u=0} \tag{A 20}$$

Example 2: The particle-scattering factor of linear monodisperse chains.

The generating functions $F_0(u)$ and $F_1(u)$ are given for the random polycondensate by Eq. (A 5) and (A 6). Eq. (A 19) yields then for Gaussian chains

$$H(u) = F_0(u) + 2\alpha u F_1(u) \sum_{n=1}^{\infty} \left(\frac{\alpha u}{F_1(u)}\right)^{n-1} \phi^n$$

$$H(u) = F_0(u) + \frac{2\alpha u \phi F_1(u)}{1 - \frac{\alpha u}{F_1(u)} \phi} = F_0(u) + \frac{2\alpha u \phi F_1(u)^2}{F_1(u) - \alpha u \phi} \tag{A 21}$$

$$H(u) = F_0(u) \left[\frac{1 - \alpha + \alpha u(1 + \phi)}{1 - \alpha + \alpha u(1 - \phi)}\right] = F_0(u) \left[1 + \frac{2\alpha u \phi}{1 - \alpha + \alpha u(1 - \phi)}\right]$$

The tedious differentiation required in Eq. (A 20), eventually leads to

$$w_x x P_x(q) = x(1-\alpha)^2 \alpha^{x-1} \left[\frac{1+\phi}{1-\phi} + \frac{2}{x} \frac{1-\phi^x}{(1-\phi)^2}\right] \tag{A 22}$$

which together with Eq. (A 8) gives

$$P_x(q) = \frac{1}{x} \left[\frac{1+\phi}{1-\phi} + \frac{2}{x} \frac{1-\phi^x}{(1-\phi)^2}\right] \tag{A 23}$$

with

$$\phi = \exp(-b^2 q^2/6)$$

Eq. (A 23) is identical with Eq. (C.12) for the monodisperse linear chain.

H. References

1. Staudinger, H.: Organische Kolloidchemie, Braunschweig: Vieweg Verlag 1950
2. Staudinger, H.: Arbeitserinnerungen, Heidelberg: Hüthig Verlag 1961
3. Kuhn, W., Kuhn, H.: Ergeb. exakt. Naturwiss. *25:* 1 (1951)
4. Kirkwood, J. G.: Macromolecules (Ed. Auer, P. L.) New York, London, Paris: Gordon and Breach 1967
5. Debye, P.: Collected Papers, New York: Interscience Publ. 1954
6. Flory, P. J.: Principles of Polymer Chemistry, Ithaca, New York: Cornell University Press 1953
7. Stockmayer, W. H.: J. Chem. Phys. *11*, 45 (1943); *12*, 125 (1944)
8. Stockmayer, W. H.: J. Chem. Phys. *18*, 58 (1950)
9. Stockmayer, W. H., Jacobson, H.: J. Chem. Phys. *18*, 1600 (1950)
10. Stockmayer, W. H.: J. Chem. Phys. *13*, 199 (1945)
11. Kramers, H. W.: J. Chem. Phys. *14*, 415 (1946)
12. Zimm, B. H.: J. Chem. Phys. *16*, 1093, 1099 (1948)
13. Zimm, B. H., Stockmayer, W. H.: J. Chem. Phys. *17*, 1301 (1949)
14. Flory, P. J.: J. Amer. Chem. Soc. *63*, 3083, 3091, 3097 (1941)
15. Flory, P. J.: J. Amer. Chem. Soc. *69*, 30 (1947)
16. Watson, H. W.: Educational Times *19*, 115 (1873)
17. Galton, F.: Educational Times *19*, 17 (1873)
18. Watson, H. W., Galton, F.: J. Anthropol. Inst. Great Britain and Ireland *4*, 138 (1874)
19. Fisher, R. A.: The Genetic Theory of Natural Selection, London: Oxford University Press 1930; 2nd edition: New-York Dover Publications 1958
20. Yule, G. U.: Phil. Trans. Roy. Soc. (London) B *213*, 21 (1924)
21. See books on Cascade Processes: for inst. Refs. 22–24
22. Harris, T. E.: The Theory of Branching Processes, Berlin–Heidelberg–New York: Springer-Verlag 1963
23. Srinivasan, S. K.: Stochastic Theory and Cascade Processes, New York–London–Amsterdam: Elsevier 1969
24. Mode, C. J.: Multiple Branching Processes, New York–London–Amsterdam: Elsevier 1971
25. Gordon, M.: Proc. Roy. Soc. (London) A *268*, 240 (1962)
26. Good, I. J.: Proc. Cambridge Phil. Soc. *51*, 240 (1955)
27. Good, I. J.: Proc. Cambridge Phil. Soc. *56*, 267 (1960)
28. Gordon, M., Malcolm, G. N., Butler, D. S.: Proc. Roy. Soc. A *295*, 29 (1966)
29. Gordon, M., Scantlebury, G. R.: Trans. Farad. Soc. *60*, 604 (1964)
30. Gordon, M., Scantlebury, G. R.: Proc. Roy. Soc. A *292*, 380 (1966)
31. Dobson, G. R., Gordon, M.: J. Chem. Phys. *41*, 2389 (1964)
32. Dobson, G. R., Gordon, M.: J. Chem. Phys. *43*, 705 (1965)
33. Gordon, M., Kuchařik, S., Ward, T. C.: Collect. Czech. Chem. Commun. *35*, 3252 (1970)
34. Kajiwara, K., Burchard, W., Gordon, M.: Br. Polymer J. *2*, 110 (1970)
35. Burchard, W.: Macromolecules *11*, 455 (1978)
36. Schmidt, M., Burchard, W.: Macromolecules *11*, 460 (1978)
37. Macosko, C. W., Miller, D. R.: Macromolecules *9*, 199 (1976)
38. Miller, D. R., Macosko, C. W.: Macromolecules *9*, 206 (1976)
39. Miller, D. R., Macosko, C. W.: Macromolecules *11*, 656 (1978)
40. Miller, D. R., Valles, E. M., Macosco, W. W.: Polym. Eng. Sci. *19*, 272 (1979)
41. Hammersley, J. M.: Proc. Cambridge Phil. Soc. *53*, 642 (1957)
42. Kirkpatrick, S.: Rev. Mod. Phys. *45*, 574 (1973)
43. Stauffer, D.: Phys. Rep. *54*, 1 (1979)
44. Stauffer, D.: Pure Appl. Chem. *53*, 1479 (1981)
45. Stauffer, D., Coniglio, A., Adam, M.: Adv. Polymer Sci. *44*, 103 (1982)
46. For more details see monographs and reviews on (static) light scattering, for inst. Ref. 47–56
47. Oster, G.: Chem. Rev. *43*, 319 (1948)
48. Doty, P., Edsall, J. T.: Adv. Protein Chem. *6*, 35 (1951)
49. Geiduschek, E. P., Holtzer, A. M.: Adv. Biol. Phys. Med. *6*, 431 (1958)

50. Van der Hulst, H. C.: Light Scattering by Small Molecules, New York: Wiley 1957
51. Kerker, M.: The Scattering of Light, New York: Academic 1966
52. McIntyre, D., Gornick, F. (Eds.): Light Scattering from Dilute Solution, New York: Gordon & Breach 1964
53. Stacey, K. A.: Light Scattering in Physical Chemistry, London: Butterworths 1965
54. Lundberg, J. L., Hardin, I. R., Tomioka, M. K.: Elastic Scattering in: Encyclopedia of Polymer Science and Technology, Wiley 1970, Vol. 12, pp. 355
55. Huglin, M. B. (Ed.): Light Scattering from Polymer Solutions: New York: Academic 1972
56. Burchard, W.: Light Scattering Techniques, in Applied Fibre Science (Ed. Happey, F.), New York: Academic 1978, Vol. 1, Chapter 10
57. Debye, P.: Ann. Physik *46*, 809 (1915)
58. Debye, P.: Phys. Z. *31*, 419 (1930)
59. Einstein, A.: Ann. Physik *33*, 1275 (1910)
60. Stockmayer, W. H., Moore, L. D., Fixman, M., Epstein, B. N.: J. Polymer Sci. *16*, 517 (1955)
61 a. Bushuk, W., Benoit, H.: Can. J. Chem. *36*, 1616 (1958)
61 b. Benoit, H., Wippler, C.: J. Chim. Physique *57*, 524 (1960); see also the English translation in Ref. 52, pp. 211
62. Chandrasekhar, S.: Rev. Mod. Phys. *15*, 3 (1943)
63. For more details on dynamic light scattering see Ref. 64–69
64. Chu, B.: Laser Light Scattering, New York: Academic 1974
65. Cummins, H. Z., Pike, E. R. (Eds): Photon Correlation Spectroscopy, New York: Plenum 1974
66. Berne, B. J., Pecora, R.: Dynamic Light Scattering, New York: Wiley 1976
67. Cummins, H. Z., Pike, E. R. (Eds): Photon Correlation Spectroscopy and Velocimetry, New York: Plenum 1977
68. Degiorgio, V., Corzi, M., Giglio, M. (Eds): Light Scattering in Liquids and Macromolecular Solutions, New York: Plenum 1980
69. Akcasu, A. Z., Benmouna, M., Han, C.: Polymer *21*, 866 (1980)
70. Brehm, G., Bloomfield, V.: Macromolecules *8*, 290 (1975)
71. Schmidt, M., Burchard, W., Ford, N. C.: Macromolecules *11*, 452 (1978)
72. Burchard, W.: Macromolecules *11*, 455 (1978)
73. Kirkwood, J. G., Riseman, J.: J. Chem. Phys. *16*, 565 (1948)
74. Fixman, M.: J. Chem. Phys. *42*, 3831 (1965)
75. Bixon, M.: J. Chem. Phys. *58*, 1459 (1973)
76. Zwanzig, R. W.: quoted by Stockmayer, W. H. in the appendix of "Molecular Fluids", Les Houches Lectures 1973, London: Gordon & Breach 1976, pp. 105
77. Koppel, D. E.: J. Chem. Phys. *57*, 4814 (1972)
78. Pusey, P. N., Koppel, D. E., Schaefer, D. W., Camerini-Otero, R. D., Koenig, S. H.: Biochemistry *13*, 952 (1974)
79. Ackerson, B. J.: J. Chem. Phys. *64*, 242 (1976)
80. Akcasu, A. Z., Gurol. H.: J. Polymer Sci. Physics. Ed. *14*, 1 (1976)
81. Kirkwood, J. G.: J. Polymer Sci. *12*, 1 (1954); see also Ref. 4
82. Burchard, W., Schmidt, M., Stockmayer, W. H.: Macromolecules *13*, 580 (1980)
83. Kajiwara, K.: Polymer *12*, 57 (1971)
84. Kajiwara, K., Burchard, W.: Polymer *22*, 1621 (1981)
85. Benmouna et al., Ref. 86, have given another more general integral representation for S (q, t) which reads

$$\Gamma = \frac{kT}{\zeta}\, \mathbf{q} \cdot \left[\frac{1}{S(q)} + \frac{\zeta}{(2\pi)^3}\int T\,(\mathbf{q} - \mathbf{k})\frac{S(k) - 1}{S(q)}\,d^3k\right] \cdot \mathbf{q}$$

The formula has, however, not yet been applied to a specific problem.
86. Benmouna, M., Akcasu, A. Z., Daoud, M.: Macromolecules *13*, 1703 (1980)
87. Debye, P.: see Ref. 52, p. 139 or Ref 5
88. Burchard, W.: Macromolecules *7*, 841 (1974)
89. Benoit, H.: J. Polymer Sci. *11*, 507 (1953)

90. Burchard, W.: Macromolecules *10*, 919 (1977)
91. Stockmayer, W. H., Fixman, M.: Ann N. Y. Acad. Sci. *57*, 334 (1953)
92. Casassa, E. F., Berry, G. C.: J. Polymer Sci. Part A2, *4*, 881 (1966)
93. Burchard, W., Kajiwara, K., Nerger, D.: J. Polymer Sci. Physics Ed. *20*, 157 (1982)
94. Burchard, W., Kajiwara, K., Nerger, D., Stockmayer, W. H.: Macromolecules, submitted (1982)
95. Burchard, W., Bantle, S., Müller, M., Reiner, A.: Pure Appl. Chem. *53*, 1519 (1981)
96. Gordon, M., Torkington, J. A.: Pure Appl. Chem. *53*, 1461 (1981); see in particular pp. 1475 ff.
97. Gordon, M.: Polymer *20*, 681 (1979)
98. Flory, P. J.: Principles of Polymer Chemistry, Ithaca: Cornell University Press 1953, Chapter 9
99. Feller, W.: An Introduction to Probabiloty Theory and Its Applications, New York: Wiley 1968, 3rd Ed. Chapter 15 & 16
100. Price, F.: J. Chem. Phys. *36*, 209 (1962)
101. Peller, L.: J. Chem. Phys. *36*, 2976 (1962)
102. Burchard, W., Schmidt, M., Stockmayer, W. H.: Macromolecules *13*, 1265 (1980)
103. De Gennes, P.-G.: Scaling Concepts in Polymer Physics, Ithaca: Cornell University Press 1979
104. Burchard, W.: Macromolecules *5*, 604 (1972)
105. Müller, M., Burchard, W.: Makromolekulare Chem. *179*, 1821 (1978)
106. Burchard, W., Ullisch, B., Wolf, Ch.: Farad. Discuss. Chem. Soc. *57*, 56 (1974)
107. Wolf, Ch., Burchard, W.: Makromolekulare Chem. *177*, 2519 (1976)
108. Erlander, S. R., French, D.: J. Polymer Sci. *20*, 7 (1956)
109. Flory, P. J.: J. Amer. Chem. Soc. *74*, 2718 (1952)
110. Flory, P. J.: Ref. 98, pp. 365
111. Burchard, W.: Ref. 90, Appendix 2
112. Meyer, K. H., Bernfeld, P.: Helv. Chim. Acta *23*, 865 (1940)
113. Meyer, K. H.: Natural and Synthetic Polymers, New York: Interscience Publ. 150, 2nd Ed., pp. 456
114. Franken, I., Burchard, W.: Br. Polymer J. *9*, 103 (1877)
115. Burchard, W., Bantle, S., Zahir, S. A.: Makromolekulare Chem. *182*, 145 (1981)
116. Ullisch, B., Burchard, W.: Makromolekulare Chem. *178*, 1403, 1427 (1977)
117. Whitney, R. S., Burchard, W.: ibid. *181*, 869 (1980)
118. Feller, W.: Ref. 99, Chapter 11
119. Kajiwara, K., Ribeiro, C. A. M.: Macromolecules *7*, 121 (1974)
120. Flory, P. J.: J. Amer. Chem. Soc. *58*, 1877 (1936)
121. Flory, P. J.: ibid. *64*, 2205 (1942)
122. Flory, P. J.: J. Chem. Phys. *12*, 425 (1944); see also Ref. 98, Chapter 9
123. Schulz, G. V.: Z. Physikal. Chem. B *43*, 25 (1939)
124. G. V. Schulz (Ref. 123) derived a similar distribution where he made, however, right from the beginning the unnecessary approximation of large degrees of polymerization; a continuous analytic function is obtained thereby which can be differentiated and integrated. Flory's derivation is exact and thus more widely applicable. Handling of discrete distributions involves nowadays no difficulties when the technique of probability generating functions is used. See Ref. 118
125. There exists some confusion on the naming of this distribution. The most probable distribution is called by statisticians the geometric distribution; see Ref. 99, pp. 268. We shall use the names either "most probable" or "Schulz-Flory" distribution.
126. See textbooks and monographs on polymerization kinetics, e.g. Refs. 98, 127 or 128
127. Henrici-Olivé, G., Olivé, S.: Polymerisation, Weinheim: Verlag Chemie, FRG 1969
128. Bamford, C. H., Barb, W. G., Jenkins, A. D., Onyon, P. F.: The Kinetics of Vinyl Polymerization by Radical Mechanisms, London: Butterworths 1958
129. Feller, W.: see Ref. 99, Chapter 12
130. Good, I. J.: Proc. Cambridge Phil. Soc. *45*, 360 (1949)
131. Dušek, K., Prins, W.: Adv. Polymer Sci. *6*, 1 (1969)
132. De Lagrange, J. L.: Mém. de l'Acad. de Berlin, *26*, (1770); Oeuvres, Vol. II, p. 25; see also Ref. 133

133. Whittacker, E. T., Watson, G. N.: A Course of Modern Analysis, Cambridge: Cambridge University Press, 1969, pp. 132
134. Good, I. J.: Proc. Roy. Soc. A 272, 54 (1963)
135. Reiner, A.: Ph. D. Thesis, Freiburg 1982
136. Gordon, M.: Colloquia Mathematica Societatis Janos Bolyai, (North Holland Publ. Co. Amsterdam) 4, 511 (1970)
137. Gordon, M., Kajiwara, K., Peniche-Covas, C. A. L., Ross-Murphy, S. B.: Makromolekulare Chem. 176, 2413 (1975)
138. Burchard, W., Kajiwara, K., Gordon, M., Kálal, J., Kennedy, J. W.: Macromolecules 6, 642 (1973)
139. The equations in Ref. 137 are more complex since the substitution of the free endgroups by stabilizing reagents have been taken into account; see Ref. 140
140. Kálal, J., Gordon, M., Devoy, C.: Makromolekulare Chem. 152, 233 (1972)
141. Burchard, W.: Macromolecules 7, 835 (1974)
142. Pfannemüller, B., Burchard, W., Franken, I.: Stärke 28, 1 (1976)
143. Pfannemüller, B., Potartz, Ch.: ibid. 29, 73 (1977)
144. Burchard, W., Kratz, I., Pfannemüller, B.: Makromolekulare Chem. 150, 63 (1971)
145. Burchard, W., Eschwey, A., Franken, I., Pfannemüller, B.: in Fibrous Biopolymers, Colston Papers No 26, London: Butterworths 1974, pp. 365
146. Franken, I.: Ph. D. Thesis, Freiburg 1974
147. In Ref. 114 the set of generating functions obtained from Eq. (C.119) after cascade substitution, were solved explicitly without making use of the matrix formalism.
148. Kratky, O., Porod, G.: Acta Physica Austriaca 2, 133 (1948)
149. Porod, G.: ibid. 2, 255 (1948)
150. Pekeris, C. L.: Phys. Rev. 71, 268 (1947)
151. Debye, P., Bueche, A. M.: J. Appl. Phys. 20, 518 (1949)
152. Kajiwara, K., Nerger, D.: unpubl. calculations. The correlation function $\gamma(r)$ was obtained by numerical integration of the Fourier transform $r\gamma(r) = (2\pi)^{-3} \int p(q) \sin(qr)\, qdq$. Introduction of the reduced radius $x = r/\langle S^2 \rangle^{1/2}$ yields $G(x)$. Further correlation functions are given in Ref. 90, 56 and 153.
153. Ross, G.: Optica Acta 15, 451 (1968)
154. Daoud, M., Cotton, P. J.: Preprint; to be publ. in J. Physique 1982
155. Benoit, H., Holtzer, A. M., Doty, P.: J. Phys. Chem. 58, 635 (1954)
156. Zimm, B. H.: J. Chem. Phys. 16, 1099 (1948)
157. Gordon, M., Kajiwara, K., Charlesby, A.: Eur. Polymer J. 11, 385 (1975)
158. Kurata, M., Fukatsu, M.: J. Chem. Phys. 41, 2934 (1961)
159. Zimm, B. H.: ibid. 16, 1093 (1948)
160. Debye, P.: J. Phys. Colloid Chem. 51, 18 (1947)
161. Berry, G. C.: J. Chem. Phys. 44, 4550 (1966)
162. Carpenter, D. K.: in Encyclopedia of Polymer Science and Technology, New York: Wiley 1970, Vol. 12, pp. 626
163. Stockmayer, W. H., Casassa, E. F.: J. Chem. Phys. 20, 1560 (1952)
164. Yamakawa, H.: ibid. 42, 1764 (1965)
165. Yamakawa, H.: Modern Theory of Polymer Solutions, New York: Harper & Row 1971
166. Guinier, A.: C. R. Hebd. Séances Acad. Sci. 204, 1115 (1937)
167. Guinier, A., Fournet, G.: Small Angle Scattering of X-rays, New York: Wiley 1955
168. Nerger, D.: Ph. D. Thesis, Freiburg 1978
169. Kratky, O., Porod, G.: J. Colloid Sci. 4, 35 (1949)
170. Kajiwara, K., Gordon, M.: J. Chem. Phys. 59, 3626 (1973)
171. Eschwey, H.: Diploma Thesis, Freiburg 1973
172. Eschwey, H., Hallensleben, M. L., Burchard, W.: Makromolekulare Chem. 173, 235 (1973)
173. Eschwey, H., Burchard, W.: Polymer 16, 180 (1975)
174. Rinaudo, M., Burchard, W.: to be publ.
175. De Gennes, G.-G.: Ref. 103, Chapter 5.2
176. Kajiwara, K.: J. Chem. Phys. 54, 296 (1971)
177. Gordon, M., Parker, T. G.: Proc. Roy. Soc. Edinburgh A 69, 13 (1971)
178. Schulthess, G. K., Benedek, G. B., De Blois, R. W.: Macromolecules 13, 939 (1980)

179. Schmidt, M., Burchard, W.: Macromolecules *14*, 370 (1981)
180. Edwards, S. F.: Proc. Roy. Soc. *88*, 265 (1966)
181. Daoud, M.: Ph. D. Thesis, Paris 1976
182. Daoud, M., Cotton, J. P., Farnoux, B., Jannink, G., Sarma, G. Benoit, H., Duplessix, R., Picot, C., de Gennes, P.-G.: Macromolecules *8*, 804 (1975)
183. Dubois-Violette, E., de Gennes, P.-G.: Physics *3*, 181 (1967)
184. De Gennes, P.-G.: Ref. 103, Chapter 6.2.2
185. Delsanti, M.: Ph. D. Thesis, Paris 1978
186. Adam, M., Delsanti, M.: Macromolecules *10*, 1229 (1977)
187. Han, C., Akcasu, A. Z.: ibid. *14*, 1080 (1981)
188. Schmidt, M., Nerger, D., Burchard, W.: Polymer *20*, 582 (1979)
189. Burchard, W., Schmidt, M.: Ber. Bunsenges. Phys. Chem. *83*, 388 (1979)
190. ter Meer, H.-U., Burchard, W.: Colloid Polymer Sci. *258*, 657 (1980)
191. ter Meer, H.-U., Burchard, W., Wunderlich, W.: Abstracts of Communications, 27th Internat. Sympos. on Macromolecules, IUPAC Strasbourg 1981, Vol. 2, pp. 695
192. Huber, K.: Diploma Thesis, Freiburg 1982
193. Weill, G., des Cloizeaux, J.: J. Physique *40*, 99 (1979)
194. Hoffmann, M., Krömer, H., Kuhn, R.: Polymeranalytik, Stuttgart: Thieme Verlag 1977, Vol. 1, Chapter 7.1 & 7.2
195. Yamakawa, H.: Ref. 165, p. 272
196. Hermans, J. J.: Rec. Trav. Chim. *63*, 219 (1944)
197. Debye, P., Bueche, A. M.: J. Chem. Phys. *16*, 373 (1948)
198. Burchard, W., Schmidt, M.: Polymer *21*, 745 (1980)
199. Schaefgen, J. R., Flory, P. J.: J. Amer. Chem. Soc. *70*, 2709 (1948). Again, this derivation is more exact than that given by G. V. Schulz, Ref. 123
200. Franken, I., Burchard, W.: Macromolecules *6*, 848 (1973)
201. The Schulz-Zimm distribution is called by statisticians the negative binomial distribution; see Ref. 99, p. 291
202. Akcasu, A. Z., Benmouna, M.: Macromolecules *11*, 1187 (1978)
203. Tanaka, G., Stockmayer, W. H.: pers. comm.; manuscript in prep.
204. Schmidt, M., Burchard, W.: Macromolecules *14*, 210 (1981)
205. Nicholson, L. K., Higgins, J. S., Hayter, J. B.: Abstracts of Communications, 27th Intern. Sympos. on Macromolecules, IUPAC Strasbourg 1981, Vol. 2, pp. 682
206. Bantle, S., Burchard, W.: Abstracts of Communications, 28th Intern. Sympos. on Macromolecules, IUPAC Amherst, 1982
207. Bantle, S.: Ph. D. Thesis, Freiburg 1982
208. Kajiwara, K., Burchard, W.: manuscript in prep.
209. Bantle, S., Schmidt, M., Burchard, W.: Macromolecules, in press (1982)
210. Oseen, C. W.: Hydrodynamik, Leipzig: Akademische Verlagsgesellschaft 1927
211. Rotne, J., Prager, S.: J. Chem. Phys. *50*, 4831 (1961)
212. Yamakawa, H.: ibid. *53*, 436 (1970)
213. Felderhof, B. U.: Physica *89 A*, 373 (1977)
214. Pusey, P. N.: pers. comm.
215. Müller, M., Burchard, W.: Int. J. Biol. Macromol. *3*, 71 (1982)
216. Jones, G., Caroline, D.: J. Chem. Phys. *37*, 187 (1979)
217. Han, C. C.: Polymer *20*, 259 (1979)
218. Stockmayer, W. H., Schmidt, M.: Pure Appl. Chem. *54*, 407 (1982)
219. Stockmayer, W. H., Burchard, W.: J. Chem. Phys. *70*, 3138 (1979)
220. Jakeman, E.: Ref. 65, pp. 75
221. Stockmayer, W. H., Schmidt, M.: to be publ.
222. Flory, P. J.: Ref. 98, p. 384
223. Page 1524 of Ref. 95 contains a number of errors. This page was unfortunately not replaced by the corrected version
224. Burchard, W., Müller, M.: Int. J. Biol. Macromol. *2*, 225 (1980)
225. Hermans, J., Hermans, J. J.: J. Phys. Chem. *62*, 1543 (1958)
226. Dušek, K.: Makromol. Chem. Suppl. *2*, 35 (1979)
227. Dušek, K., Vojta, V.: Brit. Polymer J. *9*, 164 (1977)

228. Dušek, K., Gordon, M., Ross-Murphy, S. B.: Macromolecules *11*, 236 (1978)
229. Gordon, M., Temple, W. B.: Makromol. Chem. *160*, 277 (1972)
230. Gordon, M.: pers. comm.
231. Daoud, M., Joany, F. J. F.: Preprint; to be publ. in J. Physique 1982
232. Lubensky, T. C., Isaacson, J.: Phys. Rev. *20*, 2130 (1979)
233. Rayleigh, J. W.: Proc. Roy. Soc. *A 90*, 219 (1914)
234. Tinker, D. O.: Chem. Phys. Lipids *8*, 230 (1972)
235. Oster, G., Riley, D. P.: Acta Cryst. *5*, 1 (1952)
236. Kerker, M., Kratohvil, J. P., Matijevic, E.: J. Opt. Soc. Amer. *52*, 551 (1962)
237. Neugebauer, T.: Ann. Phys. *42*, 509 (1943)
238. Kratky, O., Porod. G.: J. Colloid Sci. *4*, 35 (1949)
239. Fournet, G.: Bull. Soc. Franc. Mineral. Cryst. *74*, 39 (1951)
240. Mittelbach, P., Porod, G.: Acta Phys. Austriaca *14*, 185, 405 (1961)
241. See Ref. 52, pp. 139
242. Stevens, P. S.: Patterns in Nature. Penguin Books Ltd., Middlesex, England 1977

M. Gordon (editor)
Received June 2, 1982

Photon Correlation Spectroscopy
of Bulk Polymers

Gary D. Patterson

Bell Laboratories Murray Hill, New Jersey 07974, U.S.A.

The use of photon correlation spectroscopy to study the dynamics of concentration fluctuations in polymer solutions and gels is now well established. In bulk polymers near the glass transition there will be slowly relaxing fluctuations in density and optical anisotropy which can also be studied by this technique. In this article we review the development of the field of photon correlation spectroscopy from bulk polymers. The theory of dynamic light scattering from pure liquids is presented and applied to polymers. The important experimental considerations involved in the collection and analysis of this type of data are discussed. Most of the article focuses on the dynamics of fluctuations near the glass transition in polymers. All the published work in this area is reviewed and the results are critically discussed. The current state of the field is summarized and many suggestions for further work are presented.

A. Introduction

Dynamic light scattering is now a standard tool in the study of molecular motion[1]. When the characteristic times are in the range 10^{-6} to 100 s, photon correlation spectroscopy(PCS) is appropriate for measuring the dynamics of the fluctuations that lead to light scattering. PCS has been used extensively to study the dynamics of individual polymer molecules in dilute solution and the dynamics of concentration fluctuations in more concentrated polymer solutions and gels. In pure fluids light scattering is due primarily to fluctuations in density and optical anisotropy. Near the glass transition these fluctuations will relax slowly enough to be studied by PCS. In this article we review the application of PCS to the study of slowly relaxing fluctuations in bulk polymers near the glass transition.

Light scattering is due to fluctuations in the local dielectric tensor ε of the medium. In fluids these fluctuations are dynamic and the scattered intensity will be a function of time and the frequency spectrum of the scattered light will differ from that of the incident light. The time dependence of the total scattered intensity is analyzed by measuring the intensity autocorrelation function

$$C(t) = \frac{\langle I(t)I(0)\rangle}{\langle I \rangle^2} \tag{1}$$

where the brackets denote a long time average. The correlation function $C(t)$ is related to the relaxation function for the dielectric fluctuations $\Phi(t)$ according to

$$C(t) = 1 + f\Phi^2(t) \tag{2}$$

where

$$\Phi(t) = \frac{\langle \Delta\varepsilon(t)\Delta\varepsilon(0)\rangle}{\langle \Delta\varepsilon^2 \rangle} \tag{3}$$

and $\Delta\varepsilon(t)$ is the magnitude of the dielectric tensor fluctuation of the correct symmetry to give rise to the scattering. The coefficient f depends on the details of the scattering experiment and will be discussed at length in the section on the analysis of the data. The incident light is typically polarized either vertically(V) or horizontally(H) with respect to the scattering plane. If both the incident and the scattered light are polarized in the vertical direction(VV), the symmetry of the fluctuations is longitudinal. The HV(VH) scattering has transverse symmetry.

Soon after the development of photon correlation spectroscopy as a tool for studying dynamic light scattering it was realized[2] that this technique might be useful for studying molecular motion in bulk polymers near the glass transition. Four pioneers in the area of light scattering and molecular dynamics (D. A. Jackson and J. G. Powles from the University of Kent at Canterbury, and E. R. Pike and J. M. Vaughan from the Royal Radar Establishment at Malvern) combined to study the polymer poly(methyl methacrylate)(PMMA). This initial paper was seriously deficient in its details, but it was clearly demonstrated that there was dynamic light scattering from PMMA near its glass transition. No further work in this field was published until 1977 when two groups[3, 4] tried to improve on the early studies of PMMA. Relaxation behavior consistent with a glass transition was observed, but the two independent studies were not even in qualitative agreement. The lack of agreement and consistency in these early papers can be traced to two important problems:

(1) It is difficult to prepare bulk polymer samples which yield only intrinsic light scattering due to thermal fluctuations in density and optical anisotropy. The samples used were poorly characterized commercial materials.

(2) The actual relaxation functions observed near the glass transition are highly nonexponential. The earliest studies naively analyzed very limited data in terms of single exponential decays or two discrete exponentials.

At the same time that Jackson et al.[2] were studying the polymer PMMA, Lallemand and Ostrowsky[5] were examining the light scattering from glycerol near its glass transition. In order to fully characterize the correlation function, data had to be collected over at least four decades in time. The composite data could be described by a relaxation function based on the Cole-Davidson[6] distribution of relaxation times. Similar studies of inorganic glass forming systems[7, 8] employed the empirical relaxation function[9]

$$\Phi(t) = exp\left(-\left(\frac{t}{\tau}\right)^{\beta}\right) \qquad (4)$$

where $0 < \beta \leq 1$ is a measure of the width of the distribution of relaxation times implied by the nonexponential decay. The advantage of the relaxation function given in Eq.(4) is that it is expressed in closed form in the time domain and in fact describes the data very well. The average relaxation time is given by

$$\langle \tau \rangle = \int_{0}^{\infty} \Phi(t)dt = \frac{\tau}{\beta} \Gamma(1/\beta) \qquad (5)$$

where $\Gamma(x)$ is the gamma function. The use of the empirical relaxation function at least allows the data to be meaningfully characterized in terms of the two quantities $\langle \tau \rangle$ and β. Since the phenomenology of relaxation near the glass transition is the same for virtually all materials including polymers, the careful studies of small molecule glasses and inorganic glasses laid the foundation for the successful study of polymers.

It had been known for many years[10] that styrene could be thermally polymerized at moderate temperatures. J. R. Stevens and coworkers[11, 12] prepared very pure and optically clean monomer and produced light scattering quality samples of polystyrene. During the summer of 1977, J. R. Stevens, G. R. Alms and the author carried out an extensive study of the depolarized Rayleigh(HV) spectra observed during the thermal polymerization of styrene[13]. The final product of this research was a polystyrene sample which displayed only intrinsic depolarized scattering due to the optical anisotropy of the liquid. The polymerization was carried out in a sealed high quality quartz scattering cell. The sample was then maintained in the liquid state until a correlator could be obtained with which to carry out the studies of the dynamics of optical anisotropy fluctuations in polystyrene near the glass transition. The results of this research were first presented[14] at the Topical Conference on Atomic Scale Structure of Amorphous Solids held at IBM Yorktown Heights, New York, 3–5 April, 1978. The first full paper on this subject appeared[15] in January 1979. Independent work on polystyrene in the laboratory of A. M. Jamieson gave the same average relaxation times[16] and β parameter. Thus, the field of photon correlation spectroscopy of bulk polymers was finally established. Enough further results have now been obtained to warrant a thorough discussion of the principles and practice of this promising new area of research.

B. Theory

A full theory of light scattering from a viscoelastic medium has been presented by Rytov[17]. This theory has been extensively discussed in the context of Rayleigh-Brillouin spectroscopy of bulk polymers[18-21]. In the present paper the results of that work will be presented and extended to include relaxation near the glass transition.

Light scattering due to longitudinal density fluctuations will be considered first. The spectrum of scattered light depends on the frequency dependent moduli of compression K and shear G in the combination known as the longitudinal modulus $M = K + \frac{4}{3} G$. It also depends on the density ϱ, the temperature T, the thermal conductivity $\varkappa$, the specific heats at constant volume and pressure C_V and C_P, and the thermal expansion coefficient a_T. The wavelength of the fluctuations observed by light scattering is determined by the magnitude of the scattering vector

$$q = \frac{4 \pi n}{\lambda} \sin \frac{\Theta}{2} \tag{6}$$

where n is the refractive index of the medium, λ is the wavelength of the incident light in a vacuum and Θ is the scattering angle in the scattering plane. The fluctuation wavelength is $\lambda_F = 2 \pi/q$. The spectrum also depends explicitly on q. The spectrum can also depend on the frequency dependent dielectric derivative $Y = \delta\varepsilon/\delta\varrho$, but in general the frequency dependence of Y can be ignored. The explicit formula for the VV spectrum due to density fluctuations is

$$S_{VV}(q, \omega) = \frac{- k_b T}{2 \pi i \omega} \left[\frac{Y^2 C q^2}{\Delta} - c.c. \right] \tag{7}$$

where k_b is Boltzmann's constant, $C = \dfrac{\varrho C_V}{T} + \dfrac{\varkappa q^2}{i\omega T}$, $\Delta = (Mq^2 - \varrho\omega^2) C + K^2 a_T^2 q^2$, and c.c. denotes the complex conjugate. The spectral features are determined by the roots of the equation $\Delta = 0$. Even if the dispersion equation is too complicated to solve exactly, the spectrum can still be calculated using Eq.(7) if the frequency dependences of the moduli can be specified.

There will always be at least three features: (1) A central peak called the Rayleigh line with width $\Gamma_R \approx \varkappa q^2/\varrho C_P$, (2) Two shifted peaks called Brillouin scattering with splitting $\pm \Delta\omega_l = qV_l(q)$, where $V_l(q)$ is the longitudinal sound velocity for waves of vector magnitude q, and linewidth $\Gamma_l(q) = a_l(q)V_l(q)$, where $a_l(q)$ is the attenuation coefficient for longitudinal sound waves of wavevector magnitude q. The relaxation times associated with the Rayleigh and Brillouin lines ($\tau = 1/\Gamma$) are too short to be observed using PCS. When the frequency dependence of the moduli is taken into account another central feature is predicted called the Mountain peak[22]. The shape of the Mountain peak depends on the relaxational part of the longitudinal modulus. The modulus M can be represented as

$$M(\omega) = K_0 + \int_0^\infty \frac{M_r(\tau)i\omega\tau}{1 + i\omega\tau} d\tau \qquad (8)$$

where $K_0 = 1/\beta_T$ is the static modulus of compression equal to the inverse of the isothermal compressibility, and $M_r(\tau)$ is the distribution of longitudinal relaxation strengths associated with the relaxation time τ. The Mountain peak will then have a spectral shape given by

$$S_M(\omega) = \frac{\displaystyle\int_0^\infty \frac{M_r(\tau)\tau}{1 + (\omega\tau)^2} d\tau}{\displaystyle\int_0^\infty M_r(\tau)d\tau} \qquad (9)$$

If we define a normalized distribution of relaxation strengths $H(\tau)$ in the obvious way, the relaxation function can be calculated by taking the Fourier transform of Eq.(9) to yield

$$\Phi_M(t) = \int_0^\infty H(\tau) \exp(-t/\tau)d\tau \qquad (10)$$

Thus, the relaxation function observed by photon correlation spectroscopy will be a direct measure of the relaxational part of the longitudinal modulus.

The total intensity of light scattering due to density fluctuations in the equilibrium liquid is proportional to $\varrho k_b T/K_0$. The longitudinal modulus can be represented as a complex quantity with real part M' and imaginary part iM''. The fraction of the intensity associated with the Rayleigh peak is $\gamma - 1/\gamma$ where $\gamma = C_P/C_V$. The intensity of the Brillouin peaks is given by $k_b T/[V_l(q)^2]$. The sound velocity is given by $V_l(q) = [M'(\Delta\omega_l)/\varrho]^{1/2}$ where $M'(\Delta\omega_l)$ is understood to be the adiabatic real part of the longitudinal modulus evaluated at the frequency of the Brillouin splitting. Thus, the fraction of the total intensity due to the Brillouin peaks is $K_0/[M'(\Delta\omega_l)]$. When there is no dispersion in M, this ratio reduces to $1/\gamma$, as first predicted by Landau and Placzek[23]. The intensity associated with the Mountain peak is directly attributable to the dispersion in M. The fraction of the total is given by $[M'(\Delta\omega_l) - \gamma K_0]/[M'(\Delta\omega_l)]$. The fraction of the total scattering due to density fluctuations associated with the Mountain peak can easily be greater than ½ near the glass transition.

Optical anisotropy in liquids arises because of an asymmetric distribution of atomic polarizability. Depolarized Rayleigh scattering is observed in all liquids (and gases). In monatomic gases, this optical anisotropy can be described in terms of pairs of atoms. In dense media, an elegant theory by Madden[24] predicts the scattered spectrum due to pairs of density fluctuations. The important result of this theory in the present discussion is that there will be a central peak with width $\Gamma = 2D_0 q_{max}^2$ where D_0 is the self-diffusion coefficient and q_{max} is the value of q at the maximum of the structure factor $S(q)$ determined with x-ray or neutron scattering. This central peak is related to the relaxation of the local fluid structure on a scale of approximately one molecular unit. In the case of polymer fluids, the relevant diffusion coefficient is related to the short time motion of chain segments relative to other segments on different chains. As the glass transition is approched these motions will become slow and a relaxation function due to such proces- ses should be observed by light scattering. However, the intensity of light scattering associated with pairs of density fluctuations is low and it has not yet been demonstrated that this effect is observable near the glass transition. It has been studied extensively in a series of small chain molecules[24].

The existence of molecules often creates permanent intramolecular optical anisot- ropy. The optical anisotropy of the liquid is then due to fluctuations in the orientations of the molecules or molecular subunits. If we assign a symmetric traceless anisotropy tensor α to each molecule or molecular subunit in the scattering volume, then the relaxation function for collective optical anisotropy fluctuations can be expressed as

$$\Phi_{anis}(t) = \frac{Tr\left\langle \sum_{i,j} \hat{a}_i(t)\hat{a}_j(0) \right\rangle}{Tr\left\langle \sum_{i,j} \hat{a}_i(0)\hat{a}_j(0) \right\rangle} \tag{11}$$

where Tr denotes the trace and the sum is over all pairs of units in the scattering volume. In the simplest case the relaxation function could be described by a single exponential function with time constant τ_{ls} given by

$$\tau_{ls} = \frac{g_2}{J_2}\tau_{or} \tag{12}$$

where

$$g_2 = \frac{Tr\left\langle \sum_{i,j} \hat{a}_i(0)\hat{a}_j(0) \right\rangle}{Tr\left\langle \sum_{i} \hat{a}_i(0)\hat{a}_i(0) \right\rangle} \tag{13}$$

is the static pair correlation parameter, $J_2 \approx 1$ is the corresponding dynamic pair orienta- tion parameter, and τ_{or} is the single unit reorientation relaxation time. Although the quantity g_2 can become very large near an isotropic-nematic phase transition, it is usually near 1 in atactic polymer liquids. Even in regular chain liquids such as the n-alkanes it only achieves values between 2 and 3 above the melting point. Thus, we will ignore its effect in the following discussion. In simple fluids the single particle reorientation time has been found to be well described by[25]

$$\tau_{or} = \frac{C\eta}{T} + \tau_0 \tag{14}$$

where C depends only on the size and shape of the reorienting unit, η is the shear viscosity, and τ_0 is a phenomenological constant believed to be related to the inertial limit for reorientation. For typical small molecule liquids with viscosities in the range of 1 cP, the reorientation time is in the picosecond range. As the liquid approaches the glass transition region, the shear viscosity approaches 10^{13} P and thus the orientational relaxation time near the glass transition should approach 1000 s.

The above discussion easily motivates the notion that reorientation times will become long as the liquid is cooled towards the glass transition, but it does not explain the shape of the observed relaxation function. Part of the shear viscosity in fluids is due to coupling to molecular reorientation. This effect has been studied in detail in alkane liquids[26, 27]. At low viscosities the shear modulus can be described by

$$G(\omega) = \frac{G_{trans}i\omega\tau_{trans}}{1 + i\omega\tau_{trans}} + \frac{G_{or}i\omega\tau_{or}}{1 + i\omega\tau_{or}} \tag{15}$$

where G_{trans} and G_{or} are the relaxation strengths associated with the coupling of shear to molecular translation and rotation and τ_{trans} is the corresponding translational relaxation time. For polymer fluids, there will also be terms associated with coupling to intramolecular modes of motion and to intermolecular chain entanglements. However, the relaxation strengths of these processes are small relative to G_{trans}. The relevant viscosity for the light scattering is the local viscosity which is determined only by local translational and rotational motion. In polymer fluids there will be a strong coupling between local translational and reorientational motions. As the liquid is cooled toward the glass transition, the full viscoelasticity of the medium becomes more apparent and the shear modulus can only be described by the more general relation

$$G(\omega) = \int_{0}^{\infty} \frac{G_r(\tau)i\omega\tau}{1 + i\omega\tau} d\tau \tag{16}$$

where $G_r(\tau)$ is the distribution of relaxation strengths associated with the processes which couple to shear. The part of this distribution that is associated with coupling to reorientational processes will be observed by depolarized Rayleigh scattering. Thus, the relaxation function for optical anisotropy fluctuations is predicted to be qualitatively similar to that for local shear fluctuations as a whole near the glass transition. It should also be mentioned that anisotropy fluctuations can have longitudinal symmetry and hence will contribute to the VV spectrum along with the density fluctuations. However, as outlined above, the two relaxation functions may be very similar in form near the glass transition.

C. Experimental Considerations

C.I. Sample Preparation

As noted in the introduction, the first successful studies of PCS near the glass transition in polymers employed thermally polymerized styrene. The monomer was dried over calcium hydride and vacuum distilled directly into the scattering cell. This procedure was also successfully employed to prepare poly(methyl methacrylate)(PMMA)[28] and poly-(ethyl methacrylate)(PEMA)[29]. Although our own samples were all prepared without the addition of intentional initiator, a good sample of poly(ethyl acrylate)(PEA) was prepared by thermal polymerization in the presence of an initiator[30]. All attempts to produce a truly intrinsic sample of poly(n-butyl methacrylate)(PBMA) so far have failed[31]. The reasons for this have been investigated in detail by Champion and Liddell[32]. During the later stages of the polymerization the rate actually increases due to the inhibition of termination by pairs of polymer radicals (the socalled Trommsdorf effect). Permanent defects were created in the samples during this autoacceleration phase and led to an increase in the total scattering as the polymerization went rapidly to completion. These defects produced a marked asymmetry in the angular dependence of the total

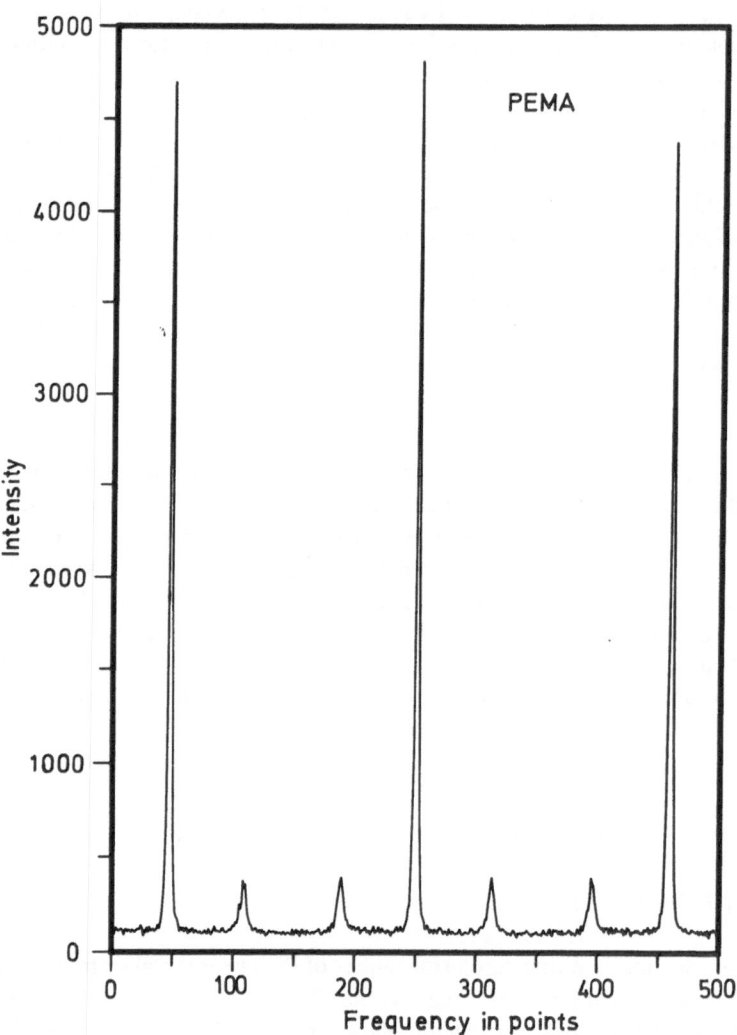

Fig. 1. Rayleigh-Brillouin spectrum of PEMA at room temperature showing several Fabry-Perot orders

scattered intensity of PBMA. Such problems are also routinely encountered in the poly-merization of PMMA[33]. The quality of the light scattering sample can be studied by examining the Rayleigh-Brillouin spectrum. A typical spectrum for PEMA in the glassy state at room temperature is shown in Fig. 1. The sample is the one made by Mahler et al.[34] and used in our study of PCS above the glass transition temperature[29]. It is our opinion that it is pointless to initiate a study of PCS in a bulk polymer until the sample quality has been verified.

Another approach to producing a good sample is to use low molecular weight poly-mer. This was the case for a thorough study of poly(propylene glycol)(PPG)[35]. The polymer can be filtered directly through $0.2\ \mu$ membrane filters into the scattering cell. One of the advantages of the very high molecular weight polymers produced by thermal polymerization is that they do not need to be contained in a cell at all. These samples do not flow under their own weight at temperatures within $60\,°C$ of T_g. The samples are removed from the cells in which they were polymerized, polished, and then annealed in a vacuum oven to relieve strains and remove any residual monomer (see discussion in Ref. 28). Although the initial studies of polystyrene in this laboratory were carried out in the sealed cell, later work has confirmed that the observed relaxation functions are independent of the container. Also, the samples studied by Jamieson[16] were free stand-ing blocks. Under these conditions no strains are introduced as the sample is cooled near T_g. These strains lead to excess elastic scattering which could distort and invalidate the results. Free standing blocks can also be studied in the glassy state. Such studies are discussed in detail below.

C.II. Light Scattering Procedure

Light scattering from bulk polymers is a very weak effect, and only part of the scattered light is due to slowly relaxing fluctuations. This means that in general a strong laser is required to carry out experiments over the full range of times allowed by the digital correlator. Typical input powers in such studies are ~ 400 mW. The limitation on input power is thermal defocussing of the beam due to absorption. Under these conditions thermal inhomogeneities are created and these defects will obscure the intrinsic scat-tering.

The low intensities also necessitate more care in eliminating sources of stray light. Well baffled optical paths are recommended. The incident light is passed through a final Glan-laser polarizer with an extinction of greater than $10^5 : 1$ when the depolarized spec-trum is measured. In order to increase the power density, the incident light is focussed to less than $200\ \mu$ beam waist in the scattering region observed by the photomultiplier. A wide variety of optical trains have been employed in the scattered light path, but there are certain important features that should be observed. The actual size of the scattering area observed should be less than the beam area, and there should be some way to determine exactly the coherence area A viewed by the detector. This point will be discussed at length below. The low intensities also affect the choice of a photomultiplier. At short delay times ($< 10^{-6}$ s), the probability of a real scattered photon per sampling interval is usually less than 0.01. This means that the probability of a correlated photo-multiplier artefact (such as an afterpulse) becomes very important. It is virtually imposs-ible to correct for these artefacts, and the usual procedure is to ignore all data points for times short enough that afterpulsing is a problem. Even though our correlator would allow a 50 ns sampling interval, it would be meaningless to use a sampling interval shorter than about 200 ns. Even at this value of Δt, the first three points would have to be ignored.

At large values of the delay time, the problem of photomultiplier dark count must be taken into account. It is important to keep the number of real counts large in comparison to the dark counts if the absolute relaxation function is to be determined. Uncorrelated dark counts increase the average number of counts per sampling interval and thereby lower the effective intercept of the measured correlation function. There is of course no guarantee that the dark counts are uncorrelated, so that it is wise just to avoid this problem. If the correlator is operated in the single-clipped mode, it is important to limit the average number of counts per sampling interval to 9. If a sampling interval of 1 s is desired, then there must be less than 1 dark count per second. This is quite possible with a good tube, but it is not routine. It is doubtful if a 15 mW laser at 6328 Å was ever intense enough to avoid this problem[16]. However, the use of full correlation instead of clipping allows higher total intensities at long sample times. This consideration alone suggests that the use of full correlation is much preferred. The recent development of a fast full correlator is especially promising for PCS in bulk polymers. It is also important to reduce the intensity due to nonthermal fluctuations such as Raman scattering. This is easily accomplished with an interference filter. In principle, the inelastic scattering such as Brillouin scattering could also be eliminated by filtering the scattered light with a Fabry-Perot interferometer, but the long term stability of such a device is not sufficient to avoid intensity fluctuations due to transmission changes. At present all the Rayleigh-Brillouin scattering must be observed and the value of the relaxation function for the scattered light will be less than 1 at the shortest observable sampling interval.

C.III. Data Analysis

The quantity that is actually measured with a digital autocorrelator is the photocount autocorrelation function

$$C(i\Delta t) = \frac{\langle n(i\Delta t)n(0)\rangle}{\langle n\rangle^2} \tag{17}$$

where i is an integer, $n(t)$ is the number of photocounts that arrived during the current sampling interval, and $\langle n\rangle$ is the average number of photocounts per sampling interval. The largest value of i that can be calculated at one time is determined by the number of discrete channels in the correlator. This number is typically between 24 and 256. Our own correlator has 96 data channels. The calculations are often further simplified by computing the single-clipped correlation function

$$C_k(i\Delta t) = \frac{\langle n(i\Delta t)n_k(0)\rangle}{\langle n\rangle\langle n_k\rangle} \tag{18}$$

where $n_k(t) = 1$ if $n > k$ and 0 otherwise, and k is called the clipping level. The use of clipping allows the calculations to be carried out using only addition (which is a fast

numerical operation) instead of multiplication. However, with the invention of fast hard-
ware multipliers, the need for clipped correlators (except at the very fastest sampling
intervals) will vanish. Most of the work on PCS from bulk polymers has been carried out
with the Malvern clipped correlator, and many laboratories still have these rugged and
reliable instruments.

The use of clipping has a profound effect on the observed correlation function and its
relation to the relaxation function. Because clipping is a nonlocal operation, the numeri-
cal factor f which relates $C_k(t)$ to $\Phi(t)$ depends explicitly on k, $\langle n \rangle$, Δt and on the
coherence area observed by the photodetector.

$$C_k(i\Delta t) = 1 + \frac{1 + k}{1 + \langle n \rangle} f(A, \Delta t, k, \langle n \rangle) \Phi^2(i\Delta t) \tag{19}$$

The effect of finite coherence area can be taken into account by measuring the value of
the apparent intercept of the relaxation function under conditions where the contribution
of Δt, k, and $\langle n \rangle$ are all equal to 1 and the fraction of the scattered light associated with
the slowly relaxing fluctuations is accurately known: We have employed dilute polymer
solutions to calibrate our system. For an ideal system, the effect of finite coherence area
can be calculated directly[36]. The effect of a finite sampling interval is to reduce the level
of correlation when Δt is comparable to the relaxation time. For a single exponential
decay, this effect can be calculated exactly[37]. When the relaxation function is highly
nonexponential, there will be many times when Δt is short relative to some of the
relaxation times and long relative to others. Under these conditions processes with $\tau < \Delta t$
make essentially no contribution to the observed correlation function after the first
channel. The full import of this situation will be explored below.

If the relaxation function were a single exponential decay, then two decades worth of
data would be sufficient to fully characterize the function and the factor f could just be
treated as an arbitrary parameter in the analysis of the data. For reasons of consistency
we choose to always correct explicitly for the effect of finite coherence area. A typical
two decade data set obtained for polystyrene is shown in Fig. 2. It took approximately
10^4 s to obtain this relaxation function. The line is the calculated best fit to the empirical
Williams-Watts function. The data is far short of the ultimate intercept at the shortest
time and far above the ultimate baseline (0.0) at the longest observed time. In order to
characterize the full relaxation function, at least four decades and preferably six decades
would be required. This requires that data sets obtained with different sampling intervals
be spliced together to obtain a composite relaxation function. To achieve good overlap
we allow the factor f to be adjusted slightly to account for the dependence on k, $\langle n \rangle$ and
Δt. We also routinely ignore the data in the first channel of the correlator at all sampling
times. In view of the known distortion of the data associated with a changing finite
sampling interval, it is very surprising (but also very gratifying) that a good superposition
is usually obtained. A composite relaxation function for PEMA covering a full eight
decades is shown in Fig. 3. Even with eight decades, the data at the shortest time is
certainly not at its intercept.

If the relaxation function were a single exponential decay, then only two parameters
would be necessary to characterize the results: a relaxation strength determined by the
fraction of the total scattered light associated with the slowly relaxing fluctuations
observed with PCS obtained from the intercept, and a relaxation time τ. With the

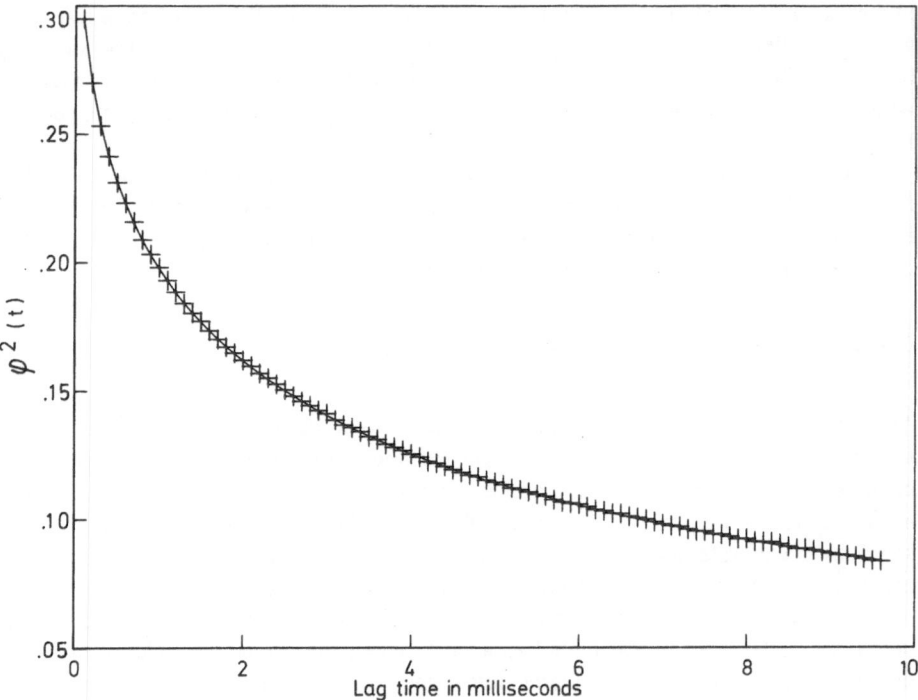

Fig. 2. Relaxation function $\Phi^2(t)$ plotted against time over two decades for polystyrene at 115 °C. The line is the best fit to the function $exp(-(t/\tau)^\beta)$

obviously nonexponential decays observed near the glass transition in bulk polymers the question arises about how much information can be reliably extracted from the data. The average relaxation time defined in Eq. (5) is a useful and meaningful parameter with which to characterize the results. Although a direct Laplace inversion to obtain the distribution of relaxation times would be desirable, it is also mathematically unstable with the type of real data obtained here. Two basic approaches are possible. The first approach is to fit the data to the empirical Williams-Watts(WW) function and obtain the parameters $\langle \tau \rangle$ and β. This has been very successful as a way to obtain the average relaxation time. The distribution of relaxation times associated with the WW function has been calculated numerically[38] as a function of the parameter β. Although there are certain mathematical artefacts associated with this distribution, the width of the distribution of relaxation times is completely determined by β and thus this parameter is a good measure of the distribution. (Although there must be an arbitrarily large number of empirical relaxation functions that could fit the data, the WW function has gained such widespread acceptance that it must be recommended for routine use.) The other approach is to assume a general distribution of relaxation times characterized by only a few parameters, and calculate the relaxation function that produces the best fit to the data. This was the procedure used in the earliest studies of glycerol[39] where the distribution of relaxation times associated with the Cole-Davidson empirical function was

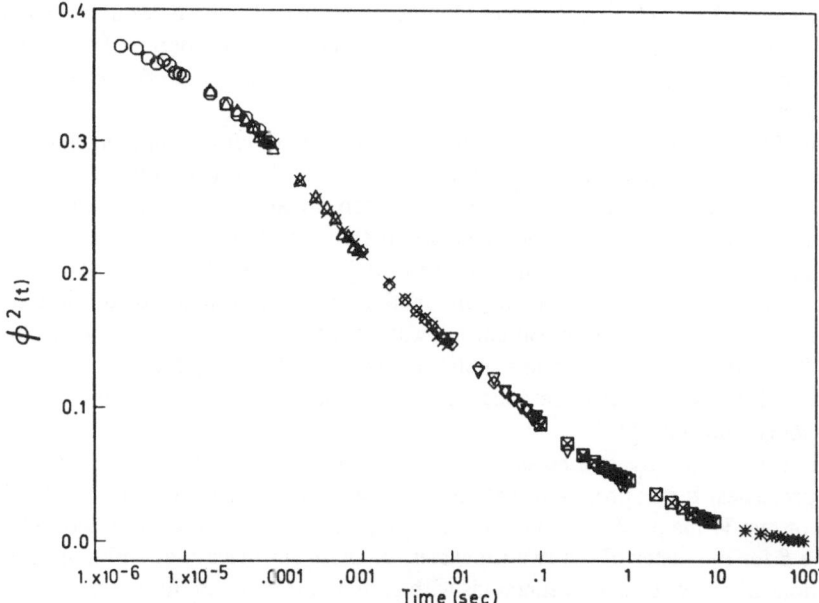

Fig. 3. Composite relaxation function $\Phi^2(t)$ obtained from six 96 point data sets plotted against log t for PEMA at 70 °C. Different symbols denote data obtained with different sampling intervals

employed. The detailed relationships between these two empirical functions have been worked out[38]. Either of the above two approaches allows the data to be characterized by the average relaxation time $\langle \tau \rangle$ and some measure of the width of a distribution of relaxation times.

There is no *a priori* reason for the observed relaxation function to be perfectly described by a WW function. It has been proposed[16] that deviations from a single WW function can be discovered by the use of a graphical technique. (Of course deviations are obvious if one examines the residual errors in the nonlinear least squares fit to the data. At least for polystyrene, there does not appear to be any systematic error in the fit to the WW function.) If the data were described perfectly by a WW function, then a plot of *log* $[- ln\Phi(t)]$ *versus* log t would be a straight line with slope equal to β and intercept given by $- \beta \log\tau$. However, the application of this graphical method to real data presents severe problems. In order to use real data in such a plot, it is necessary to renormalize the observed data so that it spans the range 0–1.0. As noted above, this is never really possible because the data is always still increasing at the shortest observable time. One could carry out a fit to a single WW function with an intercept less than 1 and use the calculated value of the intercept to renormalize the data, but this would be inconsistent. In practice[16] it appears that the renormalization factor was chosen arbitrarily, which led to a serious distortion of the data. The function *log* $[- ln]$ is singular for $\Phi(t) = 1$. If the arbitrary renormalization constant is chosen too low, the recalculated data would actually exceed 1 at the shortest times. If the time scale is restricted so that all the recalculated data lies between 0 and 1, then the plot for the shortest times will be distorted downward as the function approaches 1 more rapidly than it should. (Of course if one arbitrarily chose the constant too high, the data would never reach 1 and the plot would become horizontal at the shortest times.) Calculations which were based on perfect WW functions that were misnormalized as described above were able to reproduce the plots given by Lee et al.[16].

Because of the very low scattered intensity, the data at the shortest sampling interval is usually the poorest in quality. Arbitrary renormalization of the data followed by the graphical representation outlined above is most likely to amplify errors in the data analysis, focus attention on the inherent errors in the construction of the composite relaxation function, and give undue importance to the worst data. When the data is as limited in quality as it is for this problem, any method of analysis should be as numerically stable as possible and the maximum allowable smoothing of the data should be employed. This procedure may obscure subtle features, but only very high quality data could reliably demonstrate their presence anyway. At the present time a conservative approach seems more sensible.

The misnormalized data of Lee et al.[16] was interpreted in terms of two discrete relaxation processes. It was proposed that the relaxation function should be represented as the sum of two Williams-Watts functions. The slope at short times was claimed to be equal to the β for the faster of the two processes. Numerical calculations and graphical representations of exact relaxation functions with parameters equal to those reported by Lee et al.[16] were carried out. They did not look even qualitatively similar to their reported data. The slope at the shortest times must be related to a weighted sum of both of the β values for the sum of two WW functions. If it was desired to fit the data to a sum of two WW functions, then this could easily be carried out with a nonlinear least squares routine. In most cases it would not be possible to obtain statistically independent values of all six parameters, but at least no further errors would be introduced by faulty manipulations of the data. The graphical procedure of Lee et al.[16] cannot be recommended as of any worth in this problem.

The nonexponential decays impose limits on the shortest value of $\langle \tau \rangle$ that can be measured. If we assume that the relaxation function can be determined accurately from 2×10^{-6} s to 100 s, then the limit will be determined by the condition that the function $\Phi^2(t)$ should have a value at least as large as $1/e^2$ of its intercept value at the shortest reliable sampling interval. The value of β strongly affects the value of $\langle \tau \rangle$, but for $\beta = 0.5$, the average relaxation time is two times the value of τ. A practical limit for the shortest value of $\langle \tau \rangle$ is 10^{-5} s. The average relaxation time is determined by the longest time part of the relaxation function, so that it is probably safe to calculate $\langle \tau \rangle$ even when most of the relaxation function is not measureable, as long as the final approach to the baseline is clearly observed.

The longest value of $\langle \tau \rangle$ that can be reliably measured is determined by the longest sampling interval in the correlator times the number of channels, the dark count in the photomultiplier tube, the long term stability in the laser, and whether full correlation or clipping is employed. At present 100 s is a practical maximum for measured values of $\langle \tau \rangle$. In order to determine a relaxation time of 100 s, it is desirable to measure the correlation function for at least 1000 relaxation times. This means that run times of 10^5 s are required. This places severe requirements on the long term stability of all parts of the system. Routine measurements of $\langle \tau \rangle$ are probably better restricted to 10 s.

When the average relaxation time exceeds the longest practical sampling interval times the number of correlator channels, meaningful equilibrium data can still be obtained as long as the sample itself is at equilibrium and the run time exceeds $\langle \tau \rangle$ by a factor of 10. Data of this type will be discussed below. When the sample is deep in the glassy state, the above criteria cannot be satisfied. The discussion of the interpretation of this data will be deferred until the data obtained in the glassy state is presented.

C.IV. Comparison to Other Data

The recommended procedure outlined above yields a value of $\langle \tau \rangle$ for each measured relaxation function. Other standard relaxation techniques measure different invariants of the relaxation time distribution. In order to compare results, the relationship between the various techniques must be determined[38, 40].

One of the standard methods of analysis in PCS is to determine the average relaxation frequency $\langle \omega \rangle = \langle 1/\tau \rangle$. This is obtained by measuring the initial slope of the relaxation function. This can easily be seen from

$$\frac{d\Phi(t)}{dt} = - \int_0^\infty \frac{H(\tau)}{\tau} \, exp\,(- t/\tau)d\tau \qquad (20)$$

for $t = 0$. In actual practice the data at the shortest times is never good enough to justify taking a slope as the intercept is approached. Also, the actual value of the intercept is usually well above the value of the relaxation function at the shortest observable time. Nor is the Williams-Watts function any help in this analysis because the average frequency for the WW function happens to be infinite. This mathematical singularity has no influence on the calculation of $\langle \tau \rangle$, but it does mean that a meaningful value of $\langle 1/\tau \rangle$ cannot be calculated from the values of τ and β determined from the WW function.

The time at which the relaxation function has fallen to $1/e$ of its initial value is another arbitrary measure of the distribution of relaxation times. It is the unique parameter for a single exponential decay, but has no particular significance when there is a distribution of relaxation times.

Relaxation processes are also studied by measuring the power spectrum of the fluctuations. The power spectrum and the relaxation function are related by a Fourier transform. For a single Lorentzian peak, the normalized value of the peak is given by $S(0) = \tau/\pi$ and the frequency at which the peak falls to half its initial value is $\omega = 1/\tau$. For a distribution of relaxation times, the normalized value of the peak becomes $S(0) = \langle t \rangle/\tau$ but the half-width at half-height is no longer a unique measure of the spectrum. Nevertheless, since most spectra are characterized by this parameter, we carried out[40] an investigation of the relationship between the half-width at half-height and the parameters τ and β for the WW function. Power spectra were calculated by taking the Fourier transform of the WW function as a function of β. (The value of τ was arbitrarily set at 1, but this does not have any effect on the shape of the power spectrum.) The results are shown in Fig. 4. The half-width at half-height(hwhh) changes very rapidly with β. The product τ(hwhh) is listed in Table 1 as a function of β. The hwhh is essentially a measure of only the very longest relaxation times in the system. In practice, most spectra observed with a Fabry-Perot interferometer can be described quite well by a single Lorentzian function at least over a few half-widths. When the interferometer is set to place the hwhh

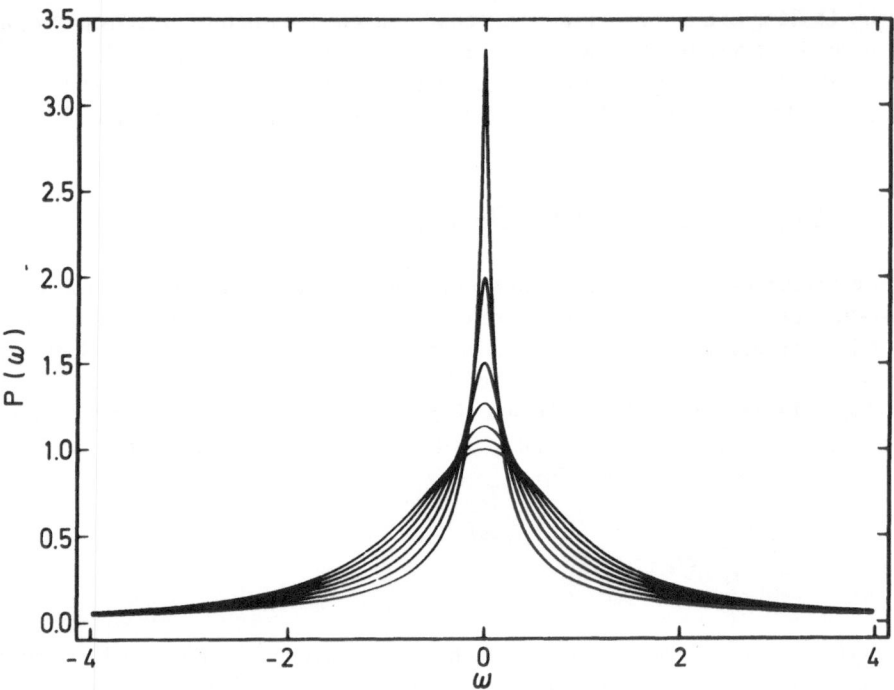

Fig. 4. Calculated normalized power spectrum $P(\omega) = S(\omega)\pi/\tau$ for different values of β ranging from 1.0 to 0.4 in increments of 0.1. The peak value of $P(\omega)$ rises as β decreases

Table 1

Relative Frequencies

β	$\tau\omega_{max}$	$\tau/\langle\tau\rangle$	τHWHH
1.0	1.0	1.0	1.0
0.9	0.943	0.950	0.871
0.8	0.886	0.883	0.725
0.7	0.833	0.790	0.563
0.6	0.787	0.665	0.391
0.5	0.743	0.500	0.224
0.4	0.659	0.300	0.088
0.3	0.639	0.108	0.015

at a reasonable fraction of the free spectral range, both low frequencies relative to hwhh and high frequencies cannot be determined with any precision. But from the above we know that the lowest frequencies in fact dominate hwhh so that it is the high frequency part of $S(\omega)$ that is obscured by the poor signal-to-noise ratio at frequencies large relative to hwhh and the overlap of multiple orders in the interferometer.

The dielectric tensor ε in a viscoelastic medium is a function of the frequency at which it is measured. It can be represented in terms of a real and imaginary part $\varepsilon(\omega) = \varepsilon'(\omega) - i\varepsilon''(\omega)$. If the frequency dependence of ε is determined by a single relaxation time, then the relationship between ε and τ is

$$\varepsilon(\omega) = \varepsilon_\infty + \frac{\varepsilon_0 - \varepsilon_\infty}{1 + i\omega\tau} \tag{21}$$

where ε_∞ is the value of the dielectric constant at frequencies substantially higher than $1/\tau$, but still much lower than the vibrational frequencies of the system, and ε_0 is the static dielectric constant. The customary[41] experiment consists of measuring $\varepsilon(\omega)$ over a wide enough frequency range to fully characterize the relaxation processes in the liquid. The normal measure of the relaxation is the frequency at which the imaginary part ε'' goes through a maximum ω_{max}. For a single relaxation time, this frequency is clearly $\omega = 1/\tau$. For a distribution of relaxation times, the frequency of maximum loss is no longer a unique measure of the relaxation. The empirical Cole-Davidson[6] function is created by replacing the relaxation denominator by $(1 + i\omega\tau)^\beta$. The average relaxation time is then $\langle \tau \rangle = \tau\beta$ where now the values of τ and β are determined from a fit of the Cole-Davidson function to the data. (It is interesting to note that for the CD function, the average frequency is also infinite. A comparison of the distributions of relaxation times associated with the CD and WW functions has been carried out[38].) The numerical relation between the parameters as determined from the WW and CD functions have been calculated[38]. Mechanical compliances can also be represented in the above fashion, and it is also customary[41] to determine the frequency of maximum mechanical loss. The relaxation function $\Phi(t)$ is related to the susceptibility χ (dielectric or mechanical) by

$$\frac{\chi(\omega) - \chi_\infty}{\chi_0 - \chi_\infty} = \int_0^\infty \exp(-i\omega t) \left(\frac{d\Phi(t)}{dt}\right) dt \tag{22}$$

In order to compare frequencies of maximum loss to $1/\langle \tau \rangle$, the frequency dependent susceptibility corresponding to the WW function was calculated as a function of β. The values of $\tau\omega_{max}$ and $\tau/\langle \tau \rangle$ are listed as a function of β in Table 1. The frequency of maximum loss correlates much better with the parameter τ and depends only weakly on β. The exact relationship presented in the Table should allow data obtained from PCS to be compared properly to dielectric or mechanical relaxation data.

D. Dynamics of the Glass Transition

D.I. Introduction

As a normal organic liquid is cooled the volume decreases and the viscosity increases. The zero shear viscosity can be represented as

$$\eta(0) = \sum_i G_i \tau_i \tag{23}$$

where the G_i are the relaxation strengths for the processes which couple to shear and the τ_i are the corresponding relaxation times. The relaxation times and relaxation strengths are both functions of temperature, but the changes in τ_i are very dramatic as the liquid is cooled towards the glass transition temperature T_g. The viscosity is observed to follow the relation

$$\eta(T) = \eta_\infty \exp\left(\frac{E}{R(T - T_2)}\right) \tag{24}$$

where η_∞ is the high temperature limiting viscosity which is typically of the order of 0.1 cP, E is an empirical constant with the units of energy per mole, R is the gas constant, and T_2 is an empirical temperature at which the equilibrium viscosity extrapolates to infinity. Equilibrium liquids can easily be obtained at temperatures well below T_g, but not below T_2.

As a liquid is cooled at a finite rate, the relaxation time spectrum will shift to longer times and a temperature region will eventually be reached where the sample is no longer in volume equilibrium. If the sample continues to be cooled at this rate it will become a glass. A glass is a nonequilibrium, mechanically unstable amorphous solid. If the sample is held at a fixed temperature near T_g the volume will relax towards its equilibrium value. In this section we will restrict our attention to equilibrium liquids at temperatures near T_g.

As mentioned above polymer liquids also have a relaxation time spectrum for intramolecular modes of motion (the socalled Rouse-Zimm modes) and for intermolecular entanglements which relax by reptation of the chains. The relaxation times for the entanglements in high polymers near the glass transition are so long that samples maintain dimensional stability if they are not subjected to external stress. Thus, high polymer liquids behave as if they were rubbers on a time scale appropriate to our experiments. It does not appear that there is any scattered intensity associated with the entanglements in polymer liquids, so that the longest observable relaxation time is associated with the internal modes of the chains.

One of the primary mechanical[42] techniques for studying the dynamics of polymer fluids near the glass transition is the measurement of the creep compliance $J(t)$. At the initial time there is a finite compliance associated with the glasslike response of the liquid J_g. The value of the compliance at this point is typically near 10^{-10} cm^2/dyne. For an uncrosslinked liquid there will be a flow term given by t/η. The total creep compliance can then be represented as

$$J(t) = J_g + J_d \psi(t) + \frac{t}{\eta} \tag{25}$$

where J_d is the recoverable compliance associated with all those processes which can store elastic energy in the fluid, and $\psi(t)$ is the retardation function for the liquid. For a high molecular weight polymer liquid, the creep compliance increases with time until a rubberlike plateau is attained. The value of the compliance in the rubbery region is largely independent of the molecular weight of the polymer as long as it is sufficiently high. The compliance in the rubbery region is typically in the range 10^{-7} to 10^{-6} cm^2/dyne. It is not experimentally easy to measure $J(t)$ at very short times, nor is it popular to wait for the entanglement plateau at temperatures near T_g. Instead the principle of time-temperature superposition is used to combine data obtained over the time interval $1–10^5$ s at different temperatures above T_g. A master curve is created at an arbitrary reference temperature by shifting the data along the time axis. The set of empirical shift factors a_T is observed to obey the same empirical equation as that for the viscosity. This means that this group of relaxation processes observed between the glassy compliance and the rubbery plateau have relaxation times which have the same temperature dependence. The collection of processes which shift according to the empirical equation

$$\tau = \tau_\infty \exp\left(\frac{E}{R(T - T_2)}\right) \tag{26}$$

is called the primary glass-rubber relaxation or α process.

In addition to the primary glass-rubber relaxation which follows the empirical shifts determined by Eq. (26), part of the recoverable compliance does not obey time-temperature superposition. The shortest time data at the lowest temperatures has a component which shifts according to the Arrhenius temperature dependence

$$\tau = \tau_\infty \exp\left(\frac{E}{RT}\right) \tag{27}$$

One set of processes which does not follow the α process shift factors is called the secondary or β relaxation.

Mechanical relaxation studies as a function of frequency have also been carried out for many polymer liquids. The range of frequencies over which one piece of dynamic apparatus can measure is very limited and the principle of time-temperature superposition is applied. Even for dielectric relaxation it is usually necessary to combine data at different temperatures in order to construct a master curve which spans a sufficient range of frequency. In this case both the real and imaginary part of the modulus or compliance are obtained but there is in fact no more information obtained than from the creep compliance. There is an exact relationship between the two representations. Nevertheless, plots of G'' or ε'' versus $\log \omega$ have the advantage that maxima are observed rather than just a monotonically increasing or decreasing function such as G' or ε'. At a temperature near T_g for an equilibrium liquid at least two maxima are usually observed corresponding to the α and β process. The frequency dependence of the loss moduli indicates that a distribution of relaxation times is associated with both processes. The width of the β process is always broader than the α when both are observed. The width of the α process appears to be constant as the temperature is changed, while the β process broadens as the temperature is lowered. The frequencies of maximum loss for the two processes approach one another as the temperature is raised above T_g. Above a certain temperature range only a single loss maximum is observed.

The longitudinal relaxation modulus observed by light scattering is dominated by processes with relaxation strengths comparable to the static modulus of compression K_0. This means that the processes which dominate the creep compliance at long times make only a small contribution to the longitudinal compliance or modulus. Only the short time processes with recoverable compliances near J_g are important here. The behaviors of the shear modulus and the bulk modulus are quite similar in this region and average relaxation times determined from volume creep experiments were observed[7] to be in good agreement with those obtained from light scattering on boron trioxide.

The processes observed in the depolarized Rayleigh spectrum correspond to internal modes of motion. Thus, they may have relaxation times which substantially exceed those obtained from the longitudinal or bulk relaxation alone. Nevertheless they are a part of the α relaxation process as it is normally observed in the creep compliance. All processes with the same shift factors make up the full α relaxation. In liquids with substantial depolarized Rayleigh scattering the slowly relaxing part of the VV scattering is also dominated by the orientation fluctuations associated with the internal modes of motion. Each internal mode contributes some intensity, but it is believed that fairly short wavelength modes dominate the scattered intensity.

The relaxation processes observed by dielectric relaxation are also related to orientational motions of the chains. The dielectric relaxation weights each mode of motion differently than the optical anisotropy relaxation, so that the two techniques are complementary. In liquids with only weak depolarized scattering, the longitudinal modulus relaxation can be observed in the VV scattering without the additional part due to anisotropy relaxation. Such results should yield identical data to that obtained directly by mechanical techniques, but it is very difficult to make such mechanical measurements and light scattering holds out the promise of being the best possible technique for studying longitudinal or bulk relaxation in liquids near the glass transition.

D.II. Temperature Studies at 1 Bar

Successful studies of the temperature dependence of PCS in bulk polymers have now been carried out for polystyrene[14-16, 43-46], PEMA[29], PMMA[28], PPG[35], and PEA[30]. As noted above, studies of PBMA have been carried out[31, 47] but the results are not yet clear. Polystyrene will be discussed first.

The relaxation function observed for polystyrene in the liquid state is dominated by anisotropy fluctuations associated with orientation fluctuations of the chain segments. Both HV and VV scattering have been observed from 100–160 °C. The average relaxation time $\langle \tau \rangle$ is the same for both polarizations. This means that anisotropy fluctuations dominate the scattering. The shape of the relaxation function is the same for all the temperatures but it does depend on the polarization. The pure HV scattering yields a β value of 0.4, while the VV scattering has a lower value (0.34). Since the VV scattering reflects both orientation fluctuations and longitudinal density fluctuations, the change in shape of the relaxation function indicates that shorter time processes are contributing to the combined VV relaxation function. This is what was expected from the discussion above. It is also worth noting that the value of the relaxation function $\Phi^2(t)$ at the short time intercept was just less than 0.4. This means that a little more than 60% of the scattered light was associated with the slowly relaxing anisotropy and longitudinal density fluctuations. The demonstration of a reasonable value for the intercept of the relaxation function is essential to any such study. Arbitrary renormalization can hide serious deficiencies in the data and even totally distort the true result. It may be more effort to calibrate the system to enable a true value of the intercept to be calculated, but in this type of study it is not safe to do otherwise.

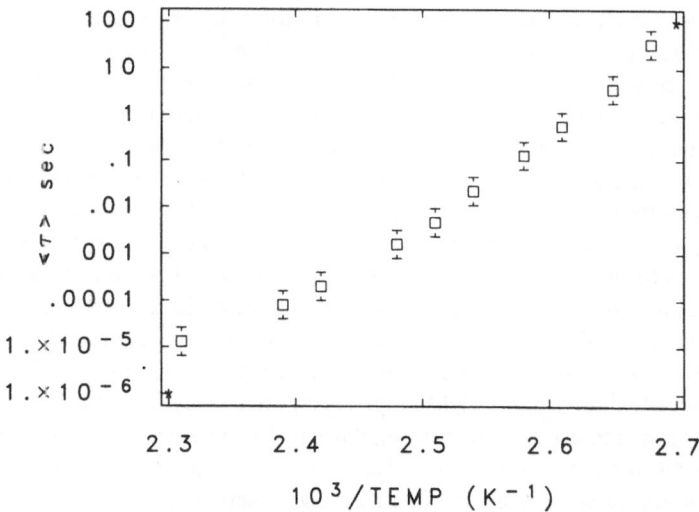

Fig. 5. Average relaxation time for polystyrene plotted logarithmically against $10^3/T$

The average relaxation time $\langle \tau \rangle$ is plotted logarithmically against the reciprocal temperature in Fig. 5. A change of nearly seven decades is observed over this temperature range. The relaxation times have been analyzed with the empirical formula given in Eq. (26) and the empirical constant $E/R = 1934$ K and $T_2 = 310$ K. These values are in very good agreement with those obtained from the temperature dependence of the viscosity by Plazek and Agarwal[48]. The temperature dependence of the dielectric loss maximum also agrees quite well[49]. The values of $\langle \tau \rangle$ are longer than those deduced from dielectric relaxation by more than a decade, but this is expected when internal modes determine the average relaxation time.

There has been a persistent claim[16, 45, 46] that an additional process is evident at short times in polystyrene. The initial evidence for this claim was the misnormalized data of Lee et al.[16]. Later analysis[45, 46] continued to show deviations from a single Williams-Watts form at the shortest times. All this work was carried out with VV polarization. Since both the longitudinal density fluctuations and the anisotropy fluctuations are observed in the VV scattering, and since they certainly have a different relaxation spectrum it is reasonable to conclude that the data should not be interpreted as a single process. However, the magnitude of the "second" process observed by Lee et al.[16, 45, 46] was very small and was claimed to be described by a single exponential relaxation function. (Our own data did not show a systematic deviation at short times and we still attribute these results to poor data analysis and the inherent error in the graphical procedure.) As noted above, the secondary relaxation process is observed to have a relaxation time distribution that is broader than the primary process. More will be said on this subject when the data in the glassy state is presented below. But it is clear that the deviations observed by Lee et al.[45, 46] should not be interpreted as the β relaxation process.

Another thoroughly studied polymer is PPG[35]. The relaxation function was studied as a function of temperature, molecular weight, and scattering angle. Only arbitrarily renormalized data was presented, but the authors were aware of the true intercept and there is no reason to suspect any problems with the results. The temperature range observed was -45 to $-65\,°C$. The average relaxation time changes by approximately four decades over this range. Both the VV and HV scattering were observed. The average relaxation time did not depend on the polarization, but for this polymer the contribution of the anisotropic scattering to VV is much lower than for polystyrene. Another major difference between the two polymers is that the optical anisotropy in polystyrene is dominated by the phenyl side chains whereas the main chain provides the chief source of anisotropy for PPG. The shape of the relaxation function was independent of temperature, but it did depend on polarization with HV giving $\beta = 0.5$ and VV giving $\beta = 0.4$. The empirical constants were $E/R = 1137\,K$ and $T_2 = 170$ K. The viscosity was also measured and observed to have similar constants. The value of T_g was constant over the molecular weight range used in this study (425–4000) so that it is not surprising that no molecular weight dependence of $\langle \tau \rangle$ was observed. Nevertheless, this is the only explicit study of this effect. (The molecular weight dependence of $\langle \tau \rangle$ in polystyrene is presently being examined over a range where T_g is changing. Different high molecular weight samples always yielded the same average relaxation time.) One of the most important results of this study was the lack of any angular dependence for $\langle \tau \rangle$. As discussed in the theory section, there should not be any q dependence to the slowly relaxing anisotropy or longitudinal density fluctuations, but it is gratifying to confirm in practice that this is so. The observation of an angular dependence to the average relaxation time[50] is a sure sign that something other than pure relaxation modes are being observed. The average relaxation times obtained by light scattering were also compared to the extensive dielectric measurements on this polymer[51]. Unfortunately, no correction for the difference between $\langle \tau \rangle$ and $1/[\omega_{max}]$ was carried out. (In fact a comment about this effect was made, but the numerical relationship was misunderstood and it was claimed that $\langle \tau \rangle$ should be less than $1/[\omega_{max}]$ for an identical distribution of relaxation times. The exact results are listed in Table 1.) When the data is corrected for this effect, there is rather good agreement between the two techniques. Such a result may be fortuitous, but for PPG, the dipole moment and the optical anisotropy are both centered in the chain backbone and thus may be reflecting very similar motions.

Similar results have been obtained for the polymer PEA[30, 52]. In this case the optical anisotropy is weaker still and only the VV scattering was studied. The temperature range was -8.3 to $26.4\,°C$ ($T_g = -24\,°C$). The average relaxation time changes by approximately four decades over this range and the empirical constants were calculated to be $E/R = 1404$ K and $T_2 = 213$ K. The shape of the relaxation function was independent of temperature with a β parameter of 0.37. Extensive dielectric data exists for this polymer[53] and the temperature dependence agrees very well. Again, the correction for different measures of the distribution of relaxation times was not applied, but in this case there is still a small difference between the two techniques when the correction is applied.

In all the above three polymers only a single process is apparently observed in the time window for PCS (10^{-6} to 100 s). The shape of the relaxation function is independent of temperature. The temperature dependence of $\langle \tau \rangle$ follows the characteristic parameters observed for mechanical or dielectric studies of the primary (α) glass-rubber relaxation. Relaxation data obtained by many techniques is collected together in the classic monograph of McCrum, Read and Williams[41]. The data is presented in the form of transition maps where the frequencies of maximum loss are plotted logarithmically

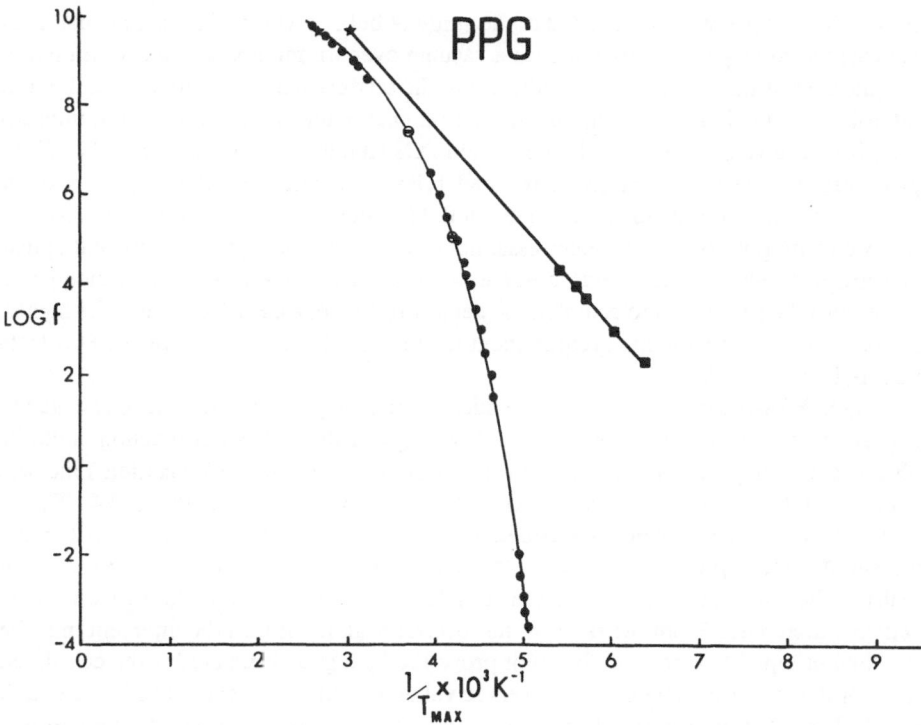

Fig. 6. Transition map for PPG showing the primary glass-rubber relaxation (α process) and the secondary relaxation (β process)

against the reciprocal temperature. A transition map for PPG[54] is shown in Fig. 6. The primary and secondary relaxation processes are clearly separated, and the frequency of maximum loss associated with the β process remains substantially higher than the observation window at all temperatures studied by Wang et al.[35]. This fact also characterizes the transition map for polystyrene and PEA[41]. There is a secondary relaxation for each of these three polymers, but it has not yet been observed by PCS. This allows the α process to be observed by itself.

As noted in the theory section, the relaxation function for the longitudinal density fluctuations should be sensitive to all processes with relaxation strengths comparable to K_0. The β relaxation in polymer fluids is often the strongest process when viewed in terms of its relaxation strength relative to K_0. The strength of the secondary process observed by Brillouin scattering[54] in PPG is definitely greater than the primary process. Thus, one should expect to observe the β process by PCS whenever it has relaxation strength in the appropriate time window. Two polymers which satisfy this criterion are PMMA[28] and PEMA[29].

Although preparation of PMMA and PEMA in optically pure form is difficult, good samples have been obtained. The first such sample was the PEMA block prepared by Mahler et al.[34]. This sample was of extremely high molecular weight and could be studied as a free standing block up to 150 °C. The procedure followed with PEMA is illustrative of the great care needed in the study of bulk polymers. The sample had to be annealed at the highest temperature in a vacuum oven for months to remove strains and volatiles. The sample was then equilibrated in the scattering oven at 150 °C until the total intensity was stable and the ratio of the central peak intensity to the Brillouin intensity stabilized at a value near 3. Only then could PCS studies be initiated. At 150 °C all the processes observed by light scattering had relaxation times less than 10^{-6} s and the observed relaxation function had a value of 0.0 for all times. Such procedures were also followed with polystyrene. It seems essential to verify that only pure density and optical anisotropy fluctuations are being observed before starting a series of experiments that can easily take six months to complete. A demonstration of a clean baseline at the highest temperature is also a minimal requirement for the initial part of any study of PCS from bulk polymers.

As the PEMA sample was cooled, evidence for slowly relaxing fluctuations began to appear at the shortest times and at 120 °C enough of the relaxation function could be observed to carry out an analysis. As noted above, the average relaxation time was approximately 10^{-5} s at this point. The value of β was observed to be 0.4. This is characteristic of the polymers described above. As the sample was cooled further the measured value of β decreased and for the relaxation function shown in Fig. 3 was equal to 0.16. The importance of obtaining the absolute value of $\Phi^2(t)$ at all times is especially demonstrated here. From the ratio of the spectral peak intensities the intercept must be less than or equal to 0.56, and this limit proved to be a good estimate for the calculated value of this quantity. The highest observed value of $\Phi^2(t)$ was near 0.4 as illustrated in Fig. 3. The function was clearly still increasing at the shortest measured time and any attempt to arbitrarily renormalize the data would seriously distort the results. The combined measurements of both the Rayleigh-Brillouin spectrum and the PCS relaxation function allow a meaningful interpretation of the true intercept. This means that it requires more than just a digital correlator to fully characterize the sample and to interpret the results. This dramatic change in the width of the relaxation function is not characteristic of a pure α relaxation. However, the average relaxation time $\langle \tau \rangle$ shifted in good accord with mechanical[55] studies of the primary glass-rubber relaxation. The empirical constants were $E/R = 2690$ K and $T_2 = 270$ K. Since the value of $\langle \tau \rangle$ is determined by the longest time part of the distribution of relaxation times, PCS can be used to study the temperature dependence of the primary glass-rubber relaxation even when the secondary process is very strong as in PEMA. The temperature dependence of the whole α process can conveniently be studied using the creep compliance or the dynamic shear compliance. The recoverable compliance associated with the β process is always very small relative to the α process. However, this is not true for the longitudinal compliance, since the intramolecular Rouse-Zimm modes make very little contribution to this quantity.

In addition to knowing the temperature shift factors, it is also necessary to know the actual value of $\langle \tau \rangle$ at some temperature. Dielectric relaxation studies often have the advantage that a frequency of maximum loss can be determined for both the primary and secondary process at the same temperature because ε'' can be measured over at least 10 decades. For PEMA there is not enough dielectric relaxation strength associated with the α process and the β process has a maximum too near in frequency to accurately resolve both processes. Only a very broad peak is observed near T_g. Studies of the frequency dependence of the shear modulus in the rubbery state could be carried out, but there have been very few such studies. One type of measurement that is commonly[56] carried out is the torsion pendulum experiment at frequencies near 1 Hz. In this experiment the temperature is varied and the real and imaginary parts of G are determined. The initial temperature is well below T_g and the experiment is continued until the value of $G'(1Hz)$ falls below 10^6 dyne/cm^2. In the region of the primary glass-rubber relaxation, the storage modulus decreases from a value near 10^9 dyne/cm^2 to a value near 10^6. Because of this dramatic logarithmic drop, it is often forgotten that the absolute drop in G' associated with the secondary processes is of the order of 10^{10} dyne/cm^2. This is further shown by examining the loss modulus G''. The value of G'' at the maximum of the β relaxation in PEMA is comparable to or greater than the value associated with the maximum of the α relaxation. Since the width of the secondary relaxation is considerably broader than the primary relaxation, the total relaxation strength is much greater for the β process in PEMA. Since at least there is a loss maximum observed by this technique for PEMA the frequency of maximum loss can be compared to that obtained by light scattering. The loss maximum at 1 Hz was observed at 78 °C. This corresponds to an average relaxation time of 0.6 s, if we assume that the appropriate distribution parameter β for the primary relaxation is near ⅓. This is in excellent agreement with the value obtained by PCS. When pure longitudinal density fluctuations are observed, it is reasonable to expect good agreement between normal mechanical relaxation studies and PCS.

It is clear from the discussion above that a large part of the relaxation function observed for PEMA is due to the β relaxation. Since the frequency of maximum loss for the primary relaxation changes much more rapidly with temperature than the secondary relaxation the total breadth of the relaxation function increases as the temperature decreases. Since the width of the β process also increases as the temperature is lowered, the relaxation function appears to be a continuous process over the whole time range. Resolution of the dielectric loss into both an α and β process at higher temperatures was achieved[57] by increasing the pressure. It has been proposed that this approach should also work for PCS[52]. However, the relaxation function follows the shape of the real part of the modulus, not the imaginary part, and even when the dielectric loss showed two maxima in PEMA the real part ε' was a continously decreasing function of frequency with no obvious separation into two parts. Nevertheless, the use of pressure in the study of PCS in bulk polymers is very important and will be discussed at length below.

One of the problems that has not yet been addressed adequately is how to extract meaningful parameters from the relaxation function that reflect the presence of the secondary relaxation process. Work on this question is being actively pursued.

A good sample of PMMA has also recently been prepared[28]. The behavior of the relaxation function as a function of temperature is very similar to that for PEMA. This was expected from the transition map which showed that the α and β processes merge at a frequency near 10^5 Hz and a temperature near 160 °C. The highest temperature at which the data could be analyzed was 150 °C. By this point the value of β was already down to 0.31. The average relaxation time was approximately 5×10^{-5} s. The value of the parameter β continued to fall and at 120 °C reached 0.15. Even with these very broad relaxation functions it was possible to obtain values of $\langle \tau \rangle$ and the temperature dependence was consistent with the measured shift factors determined by Plazek[58]. A preliminary analysis of the data yields the empirical constants $E/R = 3314$ K and $T_2 = 308$ K, but data at lower temperatures would be necessary to more fully refine these numbers. PMMA has resisted many efforts to fully characterize its dynmaics near the glass transi-

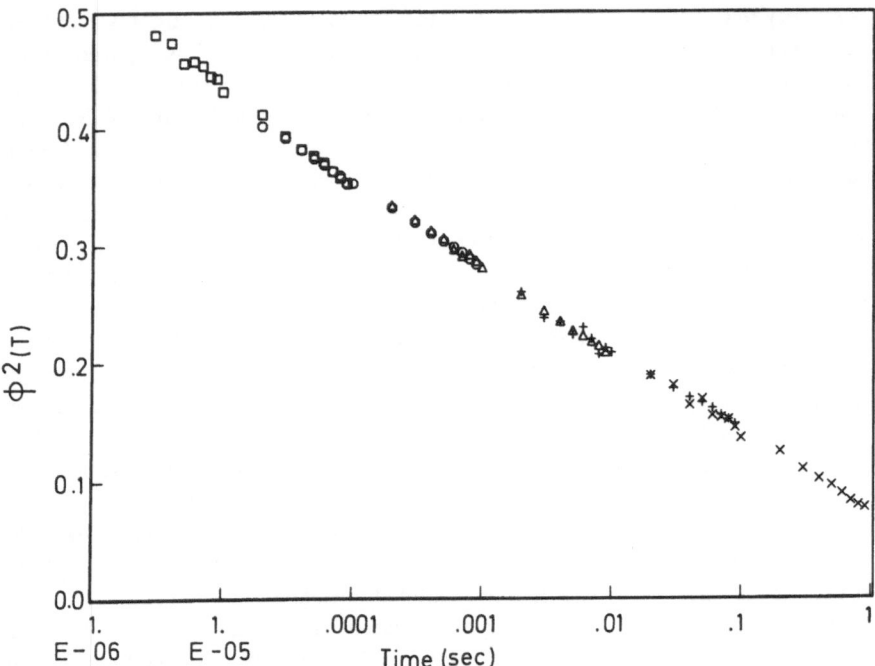

Fig. 7. Composite relaxation function for PMMA at 115 °C. The data is nearly linear over this range and has neither reached a value near the intercept at short times nor near the baseline at long times

tion[58]. The torsion pendulum results of Heijboer are in good agreement with the present results[56] as are the dielectric studies carried out by the same author[59]. The relaxation strength associated with the β process in PMMA is even larger than it is for PEMA[56]. This is supported by the very large value of the intercept observed near T_g by PCS. Data taken at 115 °C is presented in Fig. 7. The relaxation function is far below its intercept at 10^{-6} s and far above the baseline at 1 s. The composite data is almost a straight line over the entire six decades. This result helps to explain the curious and inconsistent data of all the early studies of PMMA[2-4]. With such a broad relaxation function, any two decade part looks similar. Arbitrary selection of the sampling time would lead to a decaying function with a local decay time in that interval. Ignorance of the true baseline and intercept allows almost any relaxation time to be extracted from such data. Only by obtaining the full relaxation function can one confidently interpret the results. Even then, great uncertainty may remain. This is the case for PMMA.

The published study of PBMA[31, 47] is a good illustration of the pitfalls that should be avoided in the study of PSC in bulk polymers. The most notable aspect of this work was the demonstration of a remarkable anomaly in the light scattering relaxation function over a very narrow temperature interval at 50 °C. Problems with the preparation of truly intrinsic samples of PBMA were discussed above. Even though it was known that there might be troubles with the sample, no effort was made to verify how good or bad the actual sample was.

As usual arbitrarily renormalized data was presented, but a new problem was obvious from the results. No effort was made to assure that the relaxation function had actually reached the baseline. Instead, an arbitrary baseline was introduced which forced the renormalized data to approach the value of 0.0 too rapidly. This procedure produced the result that the change in the average relaxation time between 40 and 30 °C was claimed to be approximately twice as large as between 30 and 20 °C. This would give PBMA one of the strangest glass transitions known. Since the quality of the sample was unknown, it was not clear whether the observed correlation functions had been partially heterodyned by elastic scattering from inhomogeneities in the PBMA. It was suggested that the analysis be carried out as if the data were either a pure homodyne or pure heterodyne case. However, there appears to be a series of inconsistencies between the data shown in the figures and the numbers that are reported in Table 1 of Ref. 47. If we assume that the relaxation functions are actually heterodyned, then the value of τ can be obtained by finding the $1/e$ point of the data, or by finding the time at which the ordinate of the log (ln) plot equals 0.0. For the data at 20 °C, this appears to be near 6 s, but a value near 1 s is reported. A value of 6.4 s was reported for the homodyne case, but if the observed relaxation function was truly homodyne, then the function that should have been plotted was log ($-\frac{1}{2}$ ln) to obtain the relaxation time from the point at which the ordinate reached the value 0.0.

It is possible to predict what type of results would be observed from a correct study of PBMA. The scattering is dominated by longitudinal density fluctuations, so that the relaxation function should be determined entirely by the frequency dependence of the longitudinal compliance. An estimate for the actual value of $\langle \tau \rangle$ at least at one temperature can be obtained from the torsion pendulum data of Heijboer[56]. A value near 1 s is obtained at 34 °C. Such a value is not too different from the data presented in this temperature region if we assume that the relaxation functions are heterodyned. The temperature shift factors for PBMA in the α region are only slightly smaller than those for PEMA. Thus, at 20 °C we would predict an average relaxation time near 1000 s. This is too long to be accurately studied with the correlator and arbitrary truncation of the data led to a ridiculous result at 20 °C. There is good evidence that the β relaxation is also contributing to the mechanical relaxation in this temperature interval[56]. The dielectric relaxation is completely dominated by the β process. Since the β relaxation broadens as the temperature is lowered, and since there is a big difference in the temperature dependences of the primary and secondary relaxations, the observed relaxation should broaden as the sample is cooled. The two processes merge at quite low frequencies in PBMA, and the β process is narrower than in PEMA, but there still should have been evidence for a change in shape. Perhaps the full data at 20 °C would have demonstrated this effect. A complete study of PBMA should yield rich rewards, but it must await the preparation of a truly intrinsic sample.

D.III. Pressure Studies

One of the most important variables which affects the glass transition in addition to temperature is pressure. Light scattering experiments at high pressure are far from routine, but both polystyrene[44, 60] and PEA[30, 52] have now been extensively studied. The shape of the relaxation function for polystyrene was independent of pressure as well as

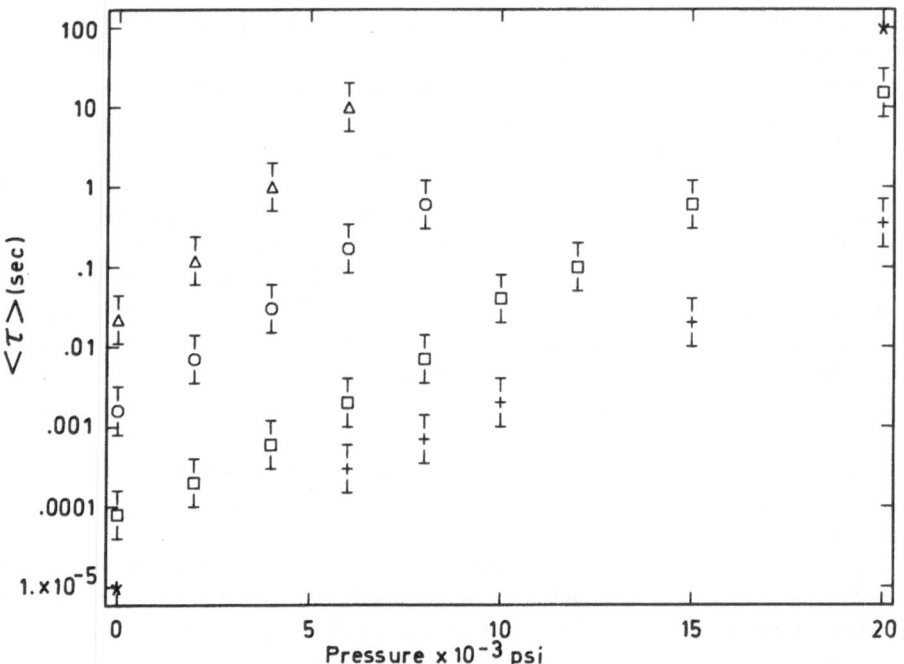

Fig. 8. Average relaxation time for polystyrene plotted logarithmically against the pressure at 120 (△), 130 (○), 146 (□) and 160 (+) °C

temperature. Average relaxation times were determined as a function of pressure at 120, 130, 146, 160 °C for polystyrene. The pressure was increased until the average relaxation time exceeded 10 s, and values as high as 1.33 kbar were reached. The apparatus could be used up to 3 kbar and at temperatures from room temperature to 200 °C. The average relaxation times $\langle \tau \rangle$ are plotted logarithmically against the pressure in Fig. 8. The logarithm of $\langle \tau \rangle$ increases linearly with pressure over this range, and the slope of the line increases as the temperature decreases. All the data can be described by the empirical equation

$$\langle \tau \rangle = \tau_\infty \exp \left(\frac{E + BP}{R(T - T_2)} \right) \tag{28}$$

where the empirical constant B can be interpreted as an activation volume for the elementary process that occurs at high temperatures. The apparent activation volume is then given by

$$\Delta V^{\neq} = \frac{BT}{T - T_2} \tag{29}$$

If we assume that T_2 is not a function of pressure, then the constant B is observed to increase slightly as the temperature is lowered with an average value near 100 cm³/mol. Such a value for B is equivalent to approximately the molar volume of one styrene subunit. An analysis of isobaric data at 0.4 kbar yielded a value of T_2 in good agreement with that determined at 1 bar. The apparent activation volume at T_g for polystyrene is equivalent to approximately six styrene subunits. The apparent activation energy is given by

$$E^* = \frac{ET^2}{(T - T_2)^2} \tag{30}$$

Thus, the apparent activation energy is changing much more rapidly near T_g than the apparent activation volume. The use of pressure greatly expands the level of information that can be gained with regard to the dynamics of the glass transition.

The thorough study of PEA by Fytas et al.[30] obtained qualitatively identical results. The parameter β was independent of pressure and temperature. The logarithm of the average relaxation time increased linearly with pressure. The parameter B was found to be 37.5 cm³/mol. This is substantially smaller than for polystyrene and reflects the much larger volume swept out by the anisotropy relaxation in polystyrene as compared to the longitudinal density relaxation in PEA. Although these types of experiments are very difficult, the information gained is well worth the effort.

As noted above, it has been suggested that pressure would be useful in separating the primary and secondary relaxations in polymers such as PEMA and PMMA. The average relaxation time associated with the α process shifts much more rapidly with pressure than the β process which often shows little or no pressure dependence. Preliminary studies[52] of PEMA have not yet been successful, but any problems should be corrected soon. One of the features of such a procedure is that the fluctuations associated with the α process do not just go away. They remain as quasi-static sources of light scattering which will partially heterodyne the remaining dynamic processes. More will be said on this subject in the section on motion in the glassy state.

D.IV. Dilution Studies

Addition of small amounts of a low molecular weight substance to a bulk polymer has a very dramatic effect on the value of T_g. Only a few percent diluent can easily lower the glass transition temperature by 100 °C. The origin of this effect is that T_g correlates with the number average molecular weight which changes very rapidly with the addition of monomer type molecules. This change in T_g is also reflected in the relaxation functions observed for bulk polymers. Only one such study has been reported[61] using PCS, but several other laboratories are pursuing this type of work and results are expected soon.

As long as the concentration of the small molecule is low ($<5\%$), the scattered intensity due to concentration fluctuations will be negligible relative to the density or anisotropy fluctuations. In polystyrene, the HV spectrum will not have any contribution due to concentration fluctuations, but in principle there could be a contribution due to the diluent anisotropy. The average relaxation time will be determined by the longest time processes and thus should reflect only the polymer fluctuations. The data were collected near the end of the thermal polymerization of styrene. Average relaxation times were determined as a function of elapsed time during the final stages of the reaction

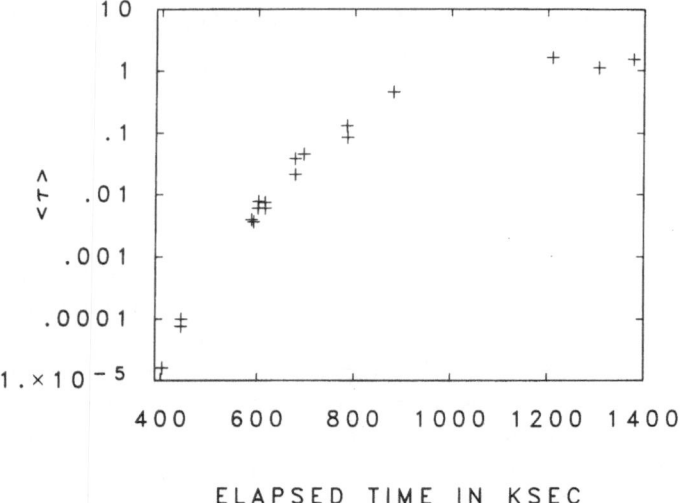

ELAPSED TIME IN KSEC

Fig. 9. Average relaxation time for polystyrene plotted logarithmically against the elapsed time during the final stages of the thermal polymerization of styrene

($>93\%$) at 90 °C until the kinetic end point was reached at this temperature. Because the sample was below the ultimate glass transition temperature the reaction did not go quite to completion. The average relaxation times $\langle \tau \rangle$ are plotted logarithmically against the elapsed time in Fig. 9. The relaxation time changes by five orders of magnitude as the reaction approaches its end point at 1 s. The shape of the relaxation function was constant during the whole experiment with the value of β equal to 0.4, the same value obtained for the bulk polymer. This result emphasizes the universality of the dynamics of the primary glass-rubber relaxation in polymers.

In some polymers, addition of very small amounts of diluent seems to fill up the holes in the liquid structure and the dynamics actually slow down. This antiplasticizer effect has not yet been examined with PCS, but this technique should prove very useful. The whole area of the effect of dilution on the PCS of bulk polymers is very promising and virtually unexplored.

E. Motion in the Glassy State

In most studies of PCS the sample is an equilibrium liquid or solution. The fluctuations which give rise to all the scattering are dynamic. If some of the fluctuations are too fast to be observed with PCS the intercept of the relaxation function is observed to be less than 1.0. The data is collected for times long relative to all the important relaxation times in the system, and thus the correlation function is stationary in time while the signal-to-noise is improved by averaging over longer times. As the average relaxation time of a polymer fluid is increased by cooling or by increasing the pressure the longest times become longer than the time window of the correlator (100 s). However, as long as the data is averaged over times long relative to the longest times, the absolute value of $\Phi(t)$ can still be obtained at the times available with the correlator. This effect is illustrated in Fig. 7 for PMMA. A practical maximum for $\langle\tau\rangle$ is thus near 10^4 s. If we arbitrarily claim that $\langle\tau\rangle$ is in the range 100–1000 s at T_g, then it is possible to obtain the true short time part of the relaxation function as much as 10 °C below T_g for many polymers.

As the average relaxation time becomes even longer than 10^4 s long times must be allowed before the sample itself achieves equilibrium. Eventually this becomes impractical and the sample becomes a glass. The longest volume relaxation times exceed the patience of the experimenter and the sample is allowed to remain in the nonequilibrium state. However, the sample does not remain in the same state as time increases because it will still relax toward the equilibrium state. The fundamental assumption of stationarity of the fluctuations is then violated and interpretation of the PCS becomes a problem. Such considerations have not stopped people from collecting data in this regime[45], but they do preclude a clear interpretation.

In addition to the intrinsic lack of stationarity, many of the fluctuations in the glass relax so slowly that they appear to be static sources of light scattering on the time scale of the data collection. These static contributions will introduce a heterodyne component into the observed relaxation function. If the fraction of the light which is quasi-static exceeds 90%, then the observed relaxation function can be interpreted as a heterodyne case and an analysis can be carried out. However, it is not clear that this limit is ever reached in practice. Only 60% of the light was slowly relaxing at all in polystyrene. If at least 90% of the slowly relaxing part becomes quasi-static the heterodyne case will still apply to the observed part of the relaxation function. For PMMA and PEMA this is unlikely to be the case at any temperature near T_g.

If the long time part of the α relaxation is not observable and the average relaxation time exceeds the run time, then the short time part of the α relaxation and perhaps the whole of the β relaxation might still be visible with the correlator. In order to examine in a systematic way the transition between the equilibrium liquid state and the glassy state we obtained at least part of the relaxation function for polystyrene as a function of pressure at 120 °C and 60 °C. At 120 °C the average relaxation time at 1 bar was 0.02 s and the pressure coefficient was approximately 6 decades per kbar. Thus average relaxation times up to 10^6 s could be obtained by applying up to 1.33 kbar of pressure. (Of course, more than a week was required to obtain equilibrium at the highest pressure, but the time was allowed and the sample was studied at equilibrium.) The glass transition pressure at 120 °C is calculated to be .67 kbar[62]. As long as the pressure was at or below this pressure the intercept was maintained at its typical value for polystyrene. A plot of $\Phi^2(t)$ against log t over the range 10^{-5} to 1 s at .67 kbar is shown in Fig. 10. There is only a slight decay over the whole time range. When the pressure was increased to 1 kbar, the intercept fell sharply to a value near 0.1. The average relaxation time at this point was near 10^4 and the typical run times were also near 10^4 s. The apparent relaxation function under these conditions is also shown in Fig. 10. When the pressure was increased to 1.33 kbar, the average relaxation time was approximately 10^6 s and the observed intercept was depressed to a value just below 0.05. The lowest line in Fig. 10 is data taken under these conditions. The effect of quasi-static fluctuations is to partially heterodyne the remaining dynamic relaxation function and thus to lower the observed intercept. Data collected at 60 °C produced even lower intercepts. In addition to determining the values of the intercept, the apparent relaxation function was determined out to 1 s. The relaxation

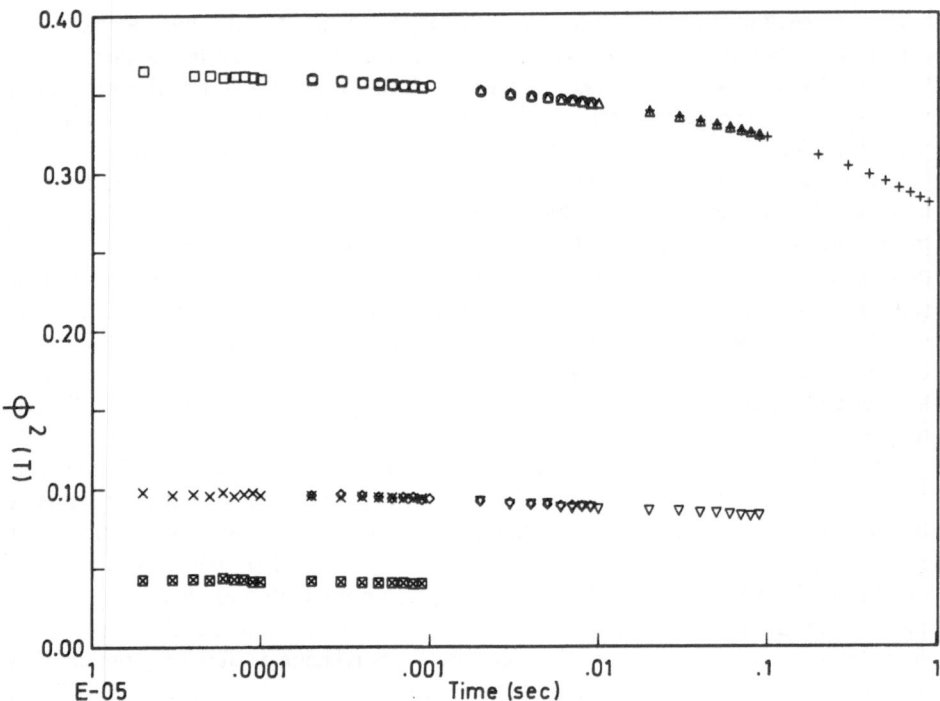

Fig. 10. Apparent relaxation function for polystyrene plotted against log t at 120 °C and pressures of 0.67, 1.0 and 1.33 kbar

function did decay over this interval, but the baseline was never reached. This demonstrates that it will be very difficult to obtain meaningful quantitative information from such data.

Data taken by Lee et al. in the glassy state for polystyrene with a 0.1 s sampling interval was arbitrarily renormalized between an arbitrary baseline and an arbitrary intercept and plotted in the usual manner[45]. It was claimed that a single relaxation time process was being observed with a relaxation time in the sampling interval. This is exactly the same type of error that characterized all the early work on PMMA[2-4]. In our work the observed apparent relaxation function for polystyrene in the glassy state decayed over the entire time window in the correlator and never reached the baseline. The observation of very broad relaxation functions in the glassy state is characteristic of all known polymers. When most of the scattered light is quasi-static it is very difficult to obtain good data, but the general qualitative features can bedetermined. Crude attempts to analyze such data[2-4, 45, 46] have not helped our understanding of the dynamics of the glassy state.

In contrast to polystyrene the observed intercepts for PMMA and PEMA in the glassy state remain high with values that are a substantial fraction of those observed in the equilibrium liquid state. Such a result should not be too surprising since it was shown above that a large part of the observed relaxation function above T_g was due to the secondary relaxation. The frequency of maximum dielectric or mechanical loss for the β

relaxation is sensitive only to the absolute temperature and not to whether the sample is a liquid or a glass. The apparent activation energy may be fairly high (30 kcal/mol) but the temperature dependence does display an Arrhenius behavior at temperatures well above and below T_g. Since the relaxation functions are so broad, substantial relaxation strength remains at short relaxation times at all temperatures near T_g in these polymers. The value of the intercept for PMMA and PEMA is still quite uncertain and the apparent relaxation functions never reach the baseline, but again there is abundant evidence of the character of the dynamic nature of the glassy state in polymers. Detailed studies of PMMA and PEMA in the glassy state have been carried out[63] and will be reported shortly.

The procedure for analyzing data obtained by PCS on the glassy state of polymers is still in a very primitive state. Much work will have to be done to determine what quantitative measures of the data will be meaningful. However, the rewards for succeeding in this problem will be very great. It is hoped that this review will stimulate further work in this area.

F. Concentration Fluctuations in Bulk Polymers

Real commercial polymers are seldom pure liquids in the sense that a chemist would define purity. Even the purest polymers are a mixture of different chain lengths. As long as the number average molecular weight is high, this type of polydispersity has no influence on PCS in bulk polymers near the glass transition. One potential source of trouble is the fact that the end groups of some commercial polymers differ substantially in polarizability from the repeating units of the chain. If the molecular weight of the polymer is relatively low, and if there is a mixture of different molecular weights, there will be light scattering due to concentration fluctuations. Another source of concentration fluctuations would be a distribution of copolymers with different fractions of each component. These types of systems were studied by Fytas et al.[50]. The polymers were poly(phenylmethyl siloxane) and a copolymer of dimethyl and phenylmethyl siloxane. The observed relaxation function was a function of the angle of observation and the decay constant varied as q^2. This was a clear sign that a diffusion process was being observed. It was shown that there really were concentration fluctuations contributing to the scattering. The relaxation functions observed for pure poly(dimethyl siloxane) showed a flat baseline. Another clear indication that the relaxation functions were not due to slowly relaxing density fluctuations near the glass transition was the very high values of the parameter β (~ 0.8). This work is a good example of the type of care that must be exercised in the study of PCS in bulk polymers. All possibilities were tested until the correct interpretation was found.

Although the above work was serendipitous, the study of concentration fluctuations in bulk polymers should be a fruitful area of research. Intentional polymer mixtures could be prepared which would allow the mutual diffusion of polymers in polymers to be obtained. Although the molecular weights might need to be kept low, the measurement of polymer motions in the bulk state would be very valuable.

G. Conclusion

The observation of slowly relaxing longitudinal density and optical anisotropy fluctuations in bulk polymers near the glass transition using PCS has now been clearly established as a valuable technique in the study of the dynamics of the glass transition. The pure optical anisotropy fluctuations associated with the rotational motions of the chain provide important complementary information to dielectric relaxation studies. When light scattering and dielectric relaxation data is combined with measurements of the shear creep compliance, a full picture of the glass-rubber relaxation in polymers can be constructed. Light scattering is sensitive to all the processes which have relaxation strength comparable to the static bulk modulus. This means that both the α and β relaxations can be detected and studied by PCS. Light scattering is already one of the most informative techniques for studying the full glass transition in the time range 10^{-6} to 100 s. Progress in the preparation of truly intrinsic light scattering samples would make this technique even more powerful.

We have examined all the published papers on the subject of PCS from bulk polymers. A clear picture has now emerged of the appropriate procedures to be followed in such studies. While only a few groups have made significant contributions to this area thus far, it is hoped that this review will stimulate new work. Photon correlation spectroscopy from bulk polymers is an especially promising field with exciting problems yet to be solved. Many of the key problems and suggestions for new work have been presented above.

H. References

 1. Berne, B. J., Pecora, R.: Dynamic light scattering. New York, Wiley 1976
 2. Jackson, D. A. et al.: J. Phys. C: Solid State Phys. 6, L55 (1973)
 3. King, T. A., Treadaway, M. F.: Chem. Phys. Lett. 50, 494 (1977)
 4. Cohen, C., Sankur, V., Pings, C. J.: J. Chem. Phys. 67, 1436 (1977)
 5. Lallemand, P., Ostrowsky, N.: Optics Comm. 8, 409 (1973)
 6. Davidson, D., Cole, R.: J. Chem. Phys. 19, 1484 (1951)
 7. Bucaro, J. A., Dardy, H. D., Corsaro, R. D.: J. Appl. Phys. 46, 741 (1975)
 8. Lai, C. C., Macedo, P. B., Montrose, C. J.: J. Amer. Ceramic Soc. 58, 120 (1975)
 9. Williams, G., Watts, D. C.: Trans. Faraday Soc. 66, 80 (1970)
10. Boundy, R. H., Boyer, R. F.: Styrene, New York, Reinhold 1962
11. Jackson, D. A., Stevens, J. R.: Mol. Phys. 30, 911 (1975)
12. Coakley R. W., et al.: J. Macromol. Sci.-Phys. 12, 511 (1976)
13. Alms, G. R., Patterson, G. D., Stevens, J. R.: J. Chem. Phys. 70, 2145 (1979)
14. Patterson, G. D.: J. Non-cryst. Sol. 31, 109 (1978)
15. Patterson, G. D., Lindsey, C. P., Stevens, J. R.: J. Chem. Phys. 70, 643 (1979)
16. Lee, H., Jamieson, A. M., Simha, R.: Macromolecules 12, 329 (1979)
17. Rytov, S. M.: Soviet Physics JETP 31, 1163 (1970)
18. Patterson, G. D., Latham, J. P.: J. Polym. Sci.: Macromol. Rev. 15, 1 (1980)
19. Patterson, G. D.: Rayleigh-Brillouin scattering in polymers, in: Methods of Experimental Physics Vol. 16A (ed.) Fava, R. A., p. 170, New York, Acad. Press 1980
20. Patterson, G. D.: CRC Crit. Rev. in Sol. State and Mat. Sci. 9, 373 (1980)

21. Patterson, G. D.: Dynamic light scattering in bulk polymers, in: Dynamic light scattering and velocimetry (ed.) Pecora, R., New York, Acad. Press 1983
22. Mountain, R. D.: J. Res. Nat. Bur. Stds. *70 A*, 207 (1966)
23. Landau, L., Placzek, G.: Phys. Z. Sowjetunion *5*, 172 (1934)
24. Patterson, G. D., Carroll, P. J.: J. Chem. Phys. *16*, 4356 (1982)
25. Alms, G. R., et al.: J. Chem. Phys. *58*, 5570 (1973)
26. Patterson, G. D., Alms, G. R.: Macromolecules *10*, 1237 (1977)
27. Patterson, G. D., Lindsey, C. P.: J. Appl. Phys. *49*, 5039 (1978)
28. Patterson, G. D., Carroll, P. J., Stevens, J. R.: J. Polym. Sci.: Polym. Phys. Ed. *20* (1982)
29. Patterson, G. D., Stevens, J. R., Lindsey, C. P.: J. Macromol. Sci.-Physics *B 18*, 641 (1980)
30. Fytas, G., et al.: Macromolecules *15*, 870 (1982)
31. Jamieson, A. M., et al.: Polymer et Eng. Sci. *21*, 965 (1981)
32. Champion, J. V., Liddell, P.: Polymer *21*, 1247 (1980)
33. Judd, R. E., Crist, B.: J. Polym. Sci.: Polym. Lett. Ed. *18*, 717 (1980)
34. Mahler, D. S. et al.: J. Appl. Phys. *49*, 5029 (1978)
35. Wang, C. H. et al.: Macromolecules *14*, 1363 (1981)
36. Jakeman, E., Oliver, C. J., Pike, E. R.: J. Phys. *A 3*, L 45 (1970)
37. Jakeman, E., Pike, E. R.: J. Phys. *A 1*, 128 (1968)
38. Lindsey, C. P., Patterson, G. D.: J. Chem. Phys. *73*, 3378 (1980)
39. Demoulin, C., Montrose, C. J., Ostrowsky, N.: Phys. Rev. *A 9*, 1740 (1974)
40. Patterson, G. D., Lindsey, C. P.: Macromolecules *14*, 83 (1981)
41. McCrum, N. G., Read, B. E., Williams, G.: Anelastic and dielectric effects in polymeric solids, London, Wiley 1967
42. Ferry, J. D.: Viscoelastic properties of polymers, New York, Wiley 1970²
43. Lindsey, C. P., Patterson, G. D., Stevens, J. R.: J. Polym. Sci.: Polym. Phys. Ed. *17*, 1547 (1979)
44. Patterson, G. D., Carroll, P. J., Stevens, J. R.: J. Polym. Sci.: Polym. Phys. Ed. *20* (1982)
45. Lee, H., Jamieson, A. M., Simha, R.: J. Macromol. Sci.-Phys. *B 18*, 649 (1980)
46. Lee, H., Jamieson, A. M., Simha, R.: Colloid and Polym. Sci. *258*, 545 (1980)
47. Jamieson, A. M. et al.: Ann. N.Y. Acad. Sci. *371*, 186 (1981)
48. Plazek, D. J., Agarwal, P.: J. Appl. Polym. Sci. *22*, 2127 (1978)
49. Kastner, S., Schlosser, E., Pohl, G.: Kolloid Z.Z. Polym. *192*, 21 (1963)
50. Fytas, G. et al.: Macromolecules *14*, 1088 (1981)
51. Yano, S. et al.: J. Polym. Sci.: Polym. Phys. Ed. *14*, 1877 (1976)
52. Fytas, G. et al.: Macromolecules *15*, 214 (1982)
53. Williams, G., Watts, D. C.: Some aspects of dielectric relaxation of amorphous polymers including the effects of a hydrostatic pressure, in: NMR Basic principles and progress Vol. 4 (ed.) Diehl, P., Fluck, E., Kosfeld R., p. 271, Berlin, Springer-Verlag 1971
54. Patterson, G. D., Douglass, D. C., Latham, J. P.: Macromolecules *11*, 263 (1978)
55. Ferry, J. D. et al.: J. Colloid Sci. *12*, 53 (1957)
56. Heijboer, J.: Mechanical properties and molecular structure of organic polymers, in: Physics of non-crystalline solids (ed.) Prins, J. A., p. 231, Amsterdam, North-Holland 1965
57. Williams, G.: Trans. Faraday Soc. *62*, 2091 (1966)
58. Plazek, D. J., Tan, V., O'Rourke, V. M.: Rheol. Acta *13*, 367 (1974)
59. Heijboer, J.: Makromol. Chem. *35 A*, 86 (1960)
60. Patterson, G. D., Stevens, J. R., Carroll, P. J.: J. Chem. Phys. *77*, 622 (1982)
61. Patterson, G. D. et al.: Macromolecules *12*, 661 (1979)
62. Coakley, R. W., Hunt, J. L., Stevens, J. R.: J. Appl. Phys. *51*, 5165 (1980)
63. Patterson, G. D., Carroll, P. J., Stevens, J. R.: to be published.

Received September 8, 1982
J. D. Ferry (editor)

Author Index Volumes 1–48

Advances in Polymer Science

Fortschritte der
Hochpolymeren-Forschung

Editors:
H.-J. Cantow, G. Dall'Asta,
K. Dušek, J.D. Ferry, H. Fujita,
M. Gordon, J.P. Kennedy,
W. Kern, S. Okamura,
C.G. Overberger, T. Saegusa,
G.V. Schulz, W.P. Slichter,
J.K. Stille

Springer-Verlag
Berlin
Heidelberg
New York

Volume 44

Polymer Networks

Editor: K. Dušek
1982. 36 figures. VII, 164 pages. ISBN 3-540-11471-8

Contents/Information:
J.E. Mark: **The Use of Model Polymer Networks to Elucidate Molecular Aspects of Rubberlike Elasticity.** The article shows that preparing elastomers by very specific chemical reactions permits an improved molecular understanding of rubberlike elasticity, and also yields materials of unusually good properties. (82 references)

S. Candau, J. Bastide, M. Delsanti: **Structural, Elastic, and Dynamic Properties of Swollen Polymer Networks.** The review reports on recent developments in the physics of gels, due to both new methods of synthesis and modern techniques for the study of microscopic properties of gels. The authors describe: recent methods for preparing labelled networks; structural properties of networks; scaling reactions for the elastic moduli of networks swollen in good solvents; and the dynamic properties of swollen networks, with special emphasis on inelastic lightscattering experiments. (134 references)

A.J. Staverman: **Properties of Phantom Networks and Real Networks.** The theory of phantom networks, initiated by James in 1947, is shown in this article to lead generally to an equation for the elastic free energy with a frontfactor. When the phantom network is in the state of lowest free energy, the Θ-state, the frontfactor equals the cycle rank, which is also equal to the number of elastic degrees of freedom.
A recent theory for real networks is shown to be inconsistent with the concept of phantom networks and an alternative theory is proposed. (37 references)

D. Stauffer, A. Coniglio, M. Adam: **Gelation and Critical Phenomena.** A detailed description of critical phenomena, critical exponents, scaling and universality for the sol-gel phase transition is given. The different predications of competing theories are listed, in particular those of "classical" theories of the Flory-Stockmayer type, and of lattice percolation theory. Experimental methods and results are reviewed to determine which theory is closest to reality. (132 references)

Volume 40

Luminescence

1981. 64 figures. V, 174 pages. ISBN 3-540-10550-6

Contents: *E.V. Anufrieva, Yu. Ya. Gotlib:* Investigation of Polymers in Solution by Polarized Luminescence. – *K.P. Ghiggino, A.J. Roberts, D. Phillips:* Time-Resolved Fluorescence Techniques in Polymer and Biopolymer Studies.

Polymers

Properties and Applications

Editorial Board: C. Brown, H.-J. Cantow,
H.J. Harwood, J.P. Kennedy, A. Ledwith,
J. Meißner, S. Okamura, G. Henrici-Olivé, S. Olivé

Volume 6
H. Janeschitz-Kriegl

Polymer Melt Rheology and Flow Birefringence

Editor: J. Meissner
1983. 144 figures. Approx. 590 pages
ISBN 3-540-11928-0

This work presents a comprehensive review of the empirical behavior of polymer melts, demonstrating for the first time the most recent molecular theories for describing this behavior. The technique of the measurement of flow birefringence is shown to be a useful tool for the investigation of rheological properties of polymer melts. The monograph is intended as an introduction into this new area of polymer science for industrial and university polymer scientists in general and rheologists and process engineers in particular. Graduate students are also addressed. The review is a fortunate combination of experimental and theoretical aspects, clearly arranged and didactically well presented.

Volume 5
J. Štepěk, H. Daoust

Additives for Plastics

1983. Approx. 54 figures. Approx. 260 pages
ISBN 3-540-90753-X

Contents: Introduction. – Additives which modify physical properties: Plasticizers. Lubricants and mold-release agents. Macromolecular modifiers. Reinforcing fillers, reinforcing agents and coupling agents. Colorants and brightening agents. Chemical and physical blowing agents. Antistatic agents. – Anti-agein additives (antidegradents): Difficultly stabilizable and nonstabilizable factors provoking plastic degradation. Heat stabilizers. Antioxidants and metal ion deactivating agents. Ultra-violet protecting agents. Flame retardants. Biocides against biological degradation of plastics. Brief survey of methods used to incorporate additives into polymer matrices.

Volume 4
A. Hebeish, T.J. Guthrie

The Chemistry and Technology of Cellulosic Copolymers

1981. 91 figures. XII, 351 pages
ISBN 3-540-10164-0

The driving force behind the great scientific interest in copolymer science and technology, is the search for products with useful, new or interesting properties. This monograph provides an informative account of new, improved cellulosic materials and the chemistry and technology involved in their production, as well as the first detailed description of grafted and modified celluloses. The information contained in this book will be of great value to researchers, manufactures, and instructors interested in the modification of cellulosics for textiles, paper printing, printing inks, paints, and packaging, as well as in polymerization processes and cellulose derivativization. (1141 references)

Volume 2
H.-H. Kausch

Polymer Fracture

1978. 180 figures, 23 tables. X, 332 pages
ISBN 3-540-08786-9

„Kausch ... is well known for his work on polymer morphology and molecular mechanics as well as his research on the strength of materials. The avowed aim of this book is to connect the more conventional statistical and continuum mechanics interpretation of fracture phenomena to the newer spectroscopic studies of highly stressed polymeric chains and the kinetics of their rupture. Relating the literature on the observed modes of viscoelasticity and irreversible deformation from polymer morphology and solid-state physics, Kausch explains the behavior and rupture of polymeric materials in terms of molecular slip and breakage processes. This leads to interesting, methodical and well-thought-out interpretations of fracture toughness, crack propagation rates and fatigue of all major polymer systems. ... Thus, the book is an outstanding contribution to our understanding of the role of chain ruptures during mechanical failure... every student and practitioner of polymer science and engineering should find this book to be a valuable resource for his work." *Physics Today*

Volume 1
B. Rånby, J. F. Rabek

ESR Spectroscopy in Polymer Research

1977. 356 figures, 29 tables. XIV, 410 pages
ISBN 3-540-08151-8

Springer-Verlag Berlin Heidelberg New York